Signals and Communication Technology

This series is devoted to fundamentals and applications of modern methods of signal processing and cutting-edge communication technologies. The main topics are information and signal theory, acoustical signal processing, image processing and multimedia systems, mobile and wireless communications, and computer and communication networks. Volumes in the series address researchers in academia and industrial R&D departments. The series is application-oriented. The level of presentation of each individual volume, however, depends on the subject and can range from practical to scientific.

Indexing: All books in "Signals and Communication Technology" are indexed by Scopus and zbMATH

For general information about this book series, comments or suggestions, please contact Mary James at mary.james@springer.com or Ramesh Nath Premnath at ramesh.premnath@springer.com.

Nour El Madhoun • Ioanna Dionysiou •
Emmanuel Bertin
Editors

Building Cybersecurity Applications with Blockchain and Smart Contracts

Springer

Editors
Nour El Madhoun
ISEP (Institut Supérieur d'Électronique de Paris)
Paris, France

Ioanna Dionysiou
Department of Computer Science
University of Nicosia
Nicosia, Cyprus

Emmanuel Bertin
Orange Innovation
Caen, France

ISSN 1860-4862 ISSN 1860-4870 (electronic)
Signals and Communication Technology
ISBN 978-3-031-50732-8 ISBN 978-3-031-50733-5 (eBook)
https://doi.org/10.1007/978-3-031-50733-5

This Springer imprint is published by the registered company Springer Nature Switzerland AG
The registered company address is: Gewerbestrasse 11, 6330 Cham, Switzerland

Paper in this product is recyclable.

Preface

Blockchain and smart contracts are significant technological advances that offer opportunities to develop potential applications in many fields. The book *Building Cybersecurity Applications with Blockchain and Smart Contracts* offers a detailed exploration of several applications in the cybersecurity field that use blockchain and smart contract technologies.

Part I introduces security in relation to blockchain and smart contracts. Chapter "Introduction to Cybersecurity Applications in Blockchain Technology" provides an overview of cybersecurity applications using these technologies.

Part II explores the application of blockchain in enhancing e-government services. Chapter "Blockchain-Integrated Secure Framework for Enhanced E-Government Services" details a framework that integrates blockchain to further secure these services.

Part III examines the use of blockchain to secure the Internet of Things (IoT). Chapter "A Blockchain-Enabled Serverless Security Mechanism for IoT-Based Drones" discusses a blockchain-based method for securing drones in an IoT context, while Chapter "Utilizing Blockchain for Safeguarding IoT-Based Robotic Networks from Spoofing Attacks" addresses the protection of IoT robotic networks also using blockchain technology.

Part IV focuses on scalability in distributed replication systems via sharding. Chapter "Sharding Distributed Replication Systems to Improve Scalability and Throughput" describes how to implement sharding in distributed replication systems to optimize scalability and throughput.

Finally, Part V examines the challenges associated with data protection in distributed ledger and blockchain technologies. Chapters "Data Protection Challenges in Distributed Ledger and Blockchain Technologies: A Combined Legal and Technical Analysis" and "Solutions to Data Protection Challenges in Distributed Ledger and Blockchain Technologies: A Combined Legal and Technical Approach" respectively offer an examination of the problems and solutions for ensuring data protection.

The book *Building Cybersecurity Applications with Blockchain and Smart Contracts* is an essential resource for those interested in the convergence of the cybersecurity field and blockchain and smart contracts technologies and who are looking to leverage these technologies to develop innovative cybersecurity applications.

Paris, France Nour El Madhoun
Nicosia, Cyprus Ioanna Dionysiou
Caen, France Emmanuel Bertin

Acknowledgments

We express our sincere gratitude to all the authors for their significant contributions to each chapter. We also thank the editors and reviewers for their proofreading work, ensuring that each chapter met high standards:

Badis Hammi, EPITA Engineering School, France
Baptiste Hemery, Orange Labs, France
Daniel Alejandro Maldonado Ruiz, Escuela Politécnica Nacional (EPN), Ecuador
Davide Frey, INRIA, France
Damien Graux, Trinity College Dublin, Ireland
Dimitri Saingre, Davidson consulting, France
El-Hacen Diallo, Université Claude Bernard Lyon 1, France
Emmanuel Bertin, Orange Labs, France
Fariba Ghaffari, IRT b-com, France
Guy Pujolle, Sorbonne Université, France
Ioanna Dionysiou, University of Nicosia, Cyprus
Jenny Torres, Escuela Politécnica Nacional (EPN), Ecuador
Kevin Atighehchi, Université Clermont Auvergne, France
Khaldoun Al Agha, Université Paris-Saclay, France
Line Guffond Le Goanvic, 360 Vision & Perspectives, France
Lina Mroueh, ISEP Engineering School, France
Lucas Sguerra, Eniblock, France
Meroua Moussaoui, Orange Labs, France
Mohamed Ali Kandi, IRIT, France
Mustapha-Kamal Benramdane, Conservatoire National des Arts et Métiers (CNAM), France
Nischal Aryal, Orange Labs, France
Nour El Madhoun, ISEP Engineering School, France
Nouredine Tamani, ISEP Engineering School, France
Rachid Guerraoui, EPFL, Switzerland
Rashid Boudour, Université Badji Mokhtar-Annaba, Algeria
Saad El Jaouhari, ISEP Engineering School, France

Samia Bouzefrane, Conservatoire National des Arts et Métiers (CNAM), France
Sara Tucci, CEA-List, France
Shyam Mohan, Sri Chandrasekarendra Saraswathi Viswa Maha Vidyalaya, India
Souvik Sengupt, IONOS, Germany
Tiphaine Henry, CEA-List, France
Tristan Bilot, Université Paris-Saclay, France
Wafa Njima, ISEP Engineering School, France
Yackolley Amoussou-Guenou, Université Paris Panthéon-Assas, France

Finally, we are deeply grateful to Mary James, Sangeetha Sankar, Olivia Ramya Chitranjan, and Brian Halm at Springer for their constant help and support throughout this project.

Contents

Part II Blockchain for Enhanced E-Government Services

Part III Securing IoT with Blockchain

Part I
Introduction to Blockchain Security

Introduction to Cybersecurity Applications in Blockchain Technology

Hanane Echchaoui, Ahlam Ferdenache, and Rachid Boudour

Abstract Cybersecurity is a compelling need for all organizations, compounded by rapid technological change and an evolving cyberattack landscape. Many areas have been explored to address the changing security requirements, and one of them is based on blockchain technology. In recent years, blockchain technology has become a top priority for many organizations. Its range of applications (smart contracts, secure transaction recording, peer-to-peer, etc.) has grown rapidly due to its inherent robustness to hackers. In this chapter, we introduce cybersecurity and present its applications in blockchain, as well as provide existing vulnerabilities and potential attacks in the state of the art to perceive the use of cybersecurity in blockchain. We also analyse the main cybersecurity challenges and how they might be addressed through advances in blockchain technology.

Keywords Cybersecurity · Blockchain · Applications · Smart contracts · Technology

1 Introduction

The rapid evolution of digital technology has transformed the way we conduct business, communicate and interact with the world. However, this digital revolution has also given rise to significant cybersecurity challenges, with cyber threats and cyberattacks becoming more sophisticated and widespread. In the quest to secure our digital assets and protect sensitive data, traditional centralized systems have shown vulnerabilities and limitations. Blockchain technology emerged as a groundbreaking solution that has the potential to revolutionize not only the way we store and exchange value but also how we approach cybersecurity. Initially introduced as

H. Echchaoui · A. Ferdenache · R. Boudour (✉)
Embedded Systems Laboratory, Badji Mokhtar-Annaba University, Annaba, Algeria
e-mail: ahlam.ferdenache@u-pec.fr

N. El Madhoun et al. (eds.), *Building Cybersecurity Applications with Blockchain and Smart Contracts*, Signals and Communication Technology,
https://doi.org/10.1007/978-3-031-50733-5_1

the underlying technology for cryptocurrencies like Bitcoin, blockchain has since evolved into a powerful tool with a myriad of applications across various industries.

At its core, blockchain is a distributed, decentralized ledger that records transactions or data in a secure, transparent and immutable manner. Unlike traditional databases that rely on a central authority to validate and store information, blockchain utilizes a network of nodes, each maintaining a copy of the entire ledger. Transactions are grouped into blocks, cryptographically linked and added to the chain in chronological order, making it nearly impossible to alter records without consensus from the network. The inherent properties of blockchain contribute to its potential as a robust cybersecurity solution. While blockchain provides enhanced security compared to centralized systems, it is not immune to cyber threats. The importance of cybersecurity in the context of blockchain cannot be overstated. As blockchain technology expands its footprint across industries, securing the underlying infrastructure and applications becomes crucial to maintaining trust, privacy and data integrity.

The primary objective of this chapter is to delve into the realm of cybersecurity applications in blockchain technology. In Sect. 2, we will point out basic elements of blockchain. In Sect. 3, we will explore how blockchain addresses security challenges present in traditional systems and examine its unique features that strengthen cybersecurity measures. In Sects. 4 and 5, we will depict applications with case studies in the real world. In Sect. 6, we present the challenges, limitations and future perspectives of Blockchain followed by a conclusion.

2 Preliminaries and Security Challenges

2.1 Basic Concepts

We begin with preliminary notions to understand the rest of the chapter:

Cybersecurity Cybersecurity is the state or process of protecting and recovering networks, devices and programs from any type of *cyberattack*. It is also known as information technology security or electronic information security. Cybersecurity applies in a variety of contexts, from business to mobile computing, and can be divided into a few common categories as illustrated in Fig. 1.

Cybersecurity application (see Fig. 1) focuses on keeping software and devices free of threats. A compromised application could provide access to the data. Successful security begins in the design stage, well before a program or device is deployed.

For readers and beginners, it is important to comprehend the basic jargon terms of this field of computing. Figure 2 presents the key terms of the realm of cybersecurity.

Vulnerability A cybersecurity vulnerability is a weakness which allows an attacker to undermine the system's data security defences. It appears at the intersection

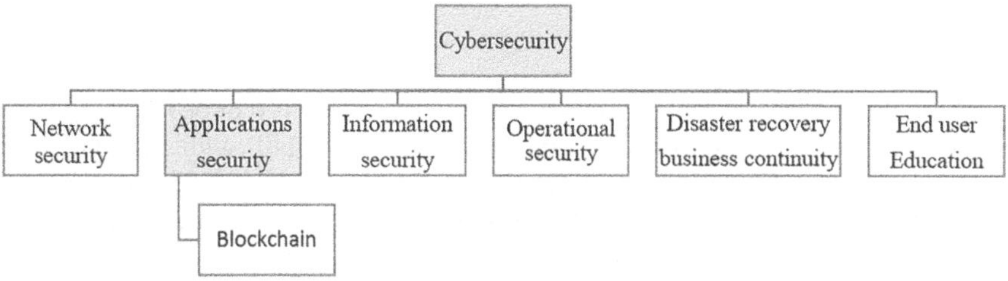

Fig. 1 Common categories of cybersecurity

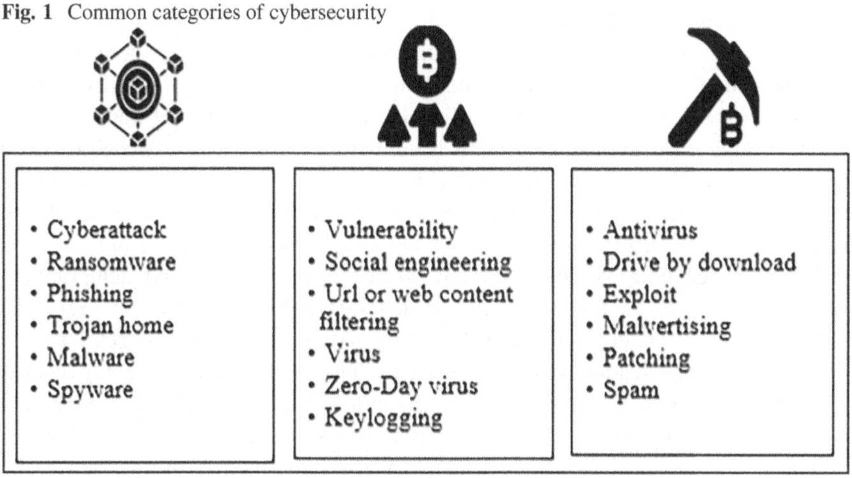

Fig. 2 Main terms of cybersecurity

of three elements: a system susceptibility or flaw, attacker access to the flaw and attacker capability to exploit the flaw load. A vulnerability is just a pretence that a cybercriminal can use to launch a full-scale attack on the system.

Cyberattack A cyberattack is classified as any type of offensive action used by cyber criminals to deploy malicious code in the system to steal, alter, destroy or take any advantage of this action. Cyberattacks can target both people and things.

We present some examples of cyberattacks and their consequences in the blockchain field. Cyberattacks are an evolving danger to organizations, employees and consumers. They may be designed to access or destroy sensitive data or extort money. *Phishing* is a way to steal personal information under the guise of reputable sources by distributing malicious links or attachments, generally by email, hoping the receiver will click, giving the attacker access to his personal information. What makes this attack very popular is the fact that attackers can target millions of potential victims at a time with such ease. A *Trojan horse* is an attack where the attacker hides a malicious program in an email. Once clicked by the victim, malware is transferred to his computer by that program, and when it's activated, it puts the victim at risk in different ways. It is sometimes used by law enforcement to lead a criminal investigation legally. For other terms, see [1].

2.2 Blockchain Technology

The simple answer to how we can determine if something is real or fake in today's world is to keep a record of it. The bank keeps a record of bills with serial numbers written on them. The Ministry of Transport in Algeria and the Department of Motor Vehicles in the USA keep a record of driver's license numbers. The relevant authority verifies the legitimacy of documents, which means that the process of verification is centralized. These centralized authorities, if corrupted, have the power to change facts or falsify records.

Nakamoto described a peer-to-peer electronic cash system called Bitcoin in [2] as the first form of money that removes the need for a central authority, where everyone keeps records of transactions and verifies the facts, so the ledger of transactions is unchangeable. Keeping money records is not the only place where decentralization can play a role. For example, Wikipedia has over 125,000 active editors to maintain its pages [3]. Each edit is public and can be verified by anyone, keeping the information decentralized and reducing the risk of not noticing if one of them goes rogue. Decentralization reduces the risk of fraud, corruption and manipulation. An innovative way to implement decentralization is blockchain technology. The Bitcoin system is the origin of blockchain technology [4]. Blockchain technology is a solution to centralization problems. It is a decentralized system of keeping records by everybody with no need for a centralized authority and maintaining a ledger that is impossible to falsify: a shared ledger with blocks of records where each block is linked to the data of the previous block. We have a chain of blocks, hence, the name blockchain. Each block comprises main data, previous hash, current hash, timestamp and other data [5]. In Fig. 3. we illustrate the general structure of a blockchain network and present the five main attributes that the blocks consist of.

Blockchain aims to establish digital network systems with four key principles: security, anonymity, transparency and immutability. It mainly works following these

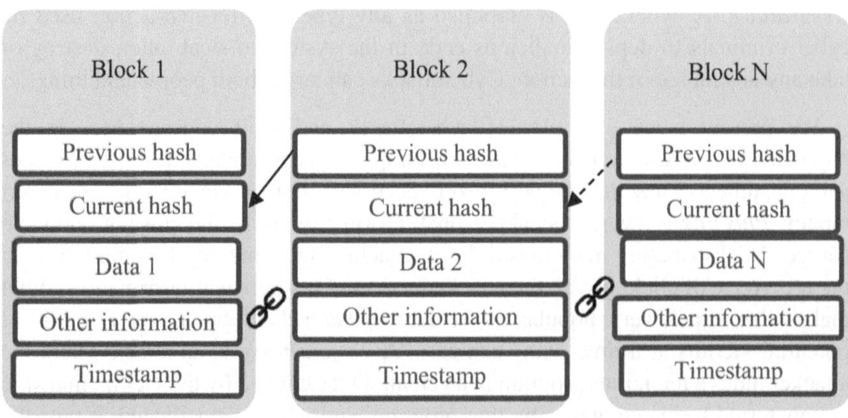

Fig. 3 Blockchain structure

steps [5]: The sending node records the new data and broadcasts it to the other nodes in the network. The receiving node checks the message from that data and stores it in a block if the message is correct. All the receiving nodes execute the consensus algorithm to the block. The new block executes the consensus algorithm and will be stored in the chain. Blockchain technology needs three elements:

Peer-to-peer network: a network of nodes that are open to everyone to communicate and share remotely.

Cryptography: the art of secure communication in a hostile environment. It allows users to verify messages and prove the authenticity of their messages. It guarantees that the communication in the first element is unaltered by malicious players.

Consensus algorithm: the rules on how to add a block to a record that needs to be agreed upon. Multiple surveys on blockchain consensus algorithms have been carried out in the literature. According to S. Bouraga in [6] the seminal blockchain consensus protocols are Proof of Work (PoW), Proof of Stake (PoS) and Practical Byzantine Fault Tolerance (PBFT). S. Pahlajani et al. in [7] proposed an overview of the main consensus algorithms in which they stated different versions of proof-based consensus as PoW, PoS, Proof of Elapsed Time, Proof of Luck, Proof of Space and their hybrid versions. S. Zhang and J. H. Lee added Delegated Proof of Stake (DPoS) and Ripple in their comparison to the list of the most used protocols in blockchain [8]. Outside of the main consensus algorithms used in blockchain, multiple new protocols have been suggested in the literature. M. Ahmed-Rengers and K. Kostiainen proposed the Robust Round Robin consensus protocol in [9] building on PoS. FastBFT is also the new protocol proposed in [10] by J. Liu et al. building on the BFT protocol. Proof-of-Learning is also a new distributed consensus protocol for validating blockchain transactions using machine learning competitions [11].

2.3 Consensus Algorithms

There are many types of consensus rules and each rule has its pros and cons. Below we present five of the well-known blockchain consensus algorithms.

Proof of Work The PoW algorithm was introduced by S. Nakasato in [2]; it is the algorithm type that the Bitcoin blockchain employs. For someone to earn the right to add a page to the ledger, they need to find a solution to a math problem. The network's computers run calculations to solve the problem consuming a lot of energy, meaning they do a lot of work, so when one of them finds the solution, they are displaying proof of work to the network which earns them the right to add the next block to the chain. We illustrate this process in Fig. 4. PoW's advantage is its high security and decentralization. However, it has several disadvantages as

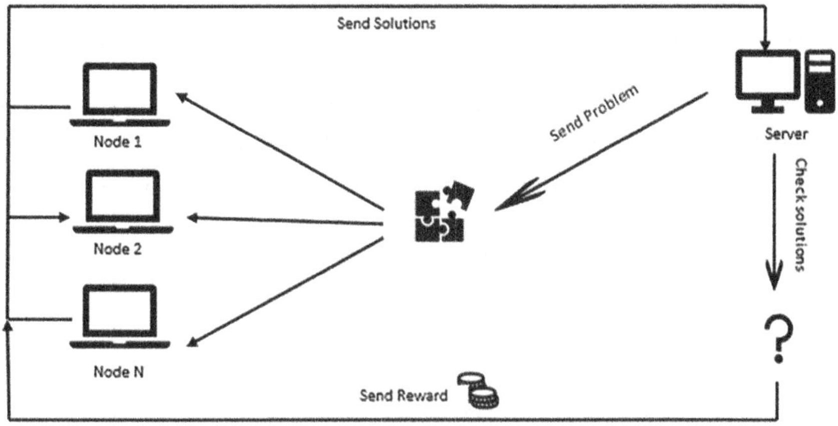

Fig. 4 Proof of Work algorithm

high block creation time, energy inefficiency, special hardware dependency, high computational cost and extensive bandwidth requirements [12].

Proof of Stake The PoS algorithm was first implemented as the cryptocurrency Peercoin in 2012 [13]. It uses an election process to randomly select one node to validate the next block. A node deposits some coins into the network as a stake to become a validator. PoS retains the advantages of PoW and overcomes some of its weaknesses. It uses significantly less energy than PoW and it is less expensive. However, it strongly depends on nodes that have the most stake, making the blockchain centralized [12].

Proof of Importance PoI is a blockchain consensus algorithm introduced by NEM (New Economic Movement) network that uses a cryptocurrency called XEM [13]. To be eligible to verify the transaction, the node needs to lock up some amount of currency similar to the PoS algorithm. But the chances for a node of verifying a transaction are not solely dependent on how much it has on stake; each node is assigned an importance score calculated from several factors (the number of vested coins, the number of transaction partners and net transactions undertaken in the past 30 days). PoI algorithm is fast and power efficient, decentralized and safe. It does not need any special hardware to mine [12].

Proof of Activity PoA was introduced by Bentov et al. in 2014 [12]. It combines the features of Proof of Work and Proof of Stake into a new approach to generate new blocks. Its mining process is similar to that of the PoW one. Miners try to solve complex mathematical problems using computing power. After the block is mined, the system is directed to PoS. Validators are randomly.

Punishment and reward A system derived from game theory, it makes sure that it is in people's best interest to always follow the rules. It is the element that

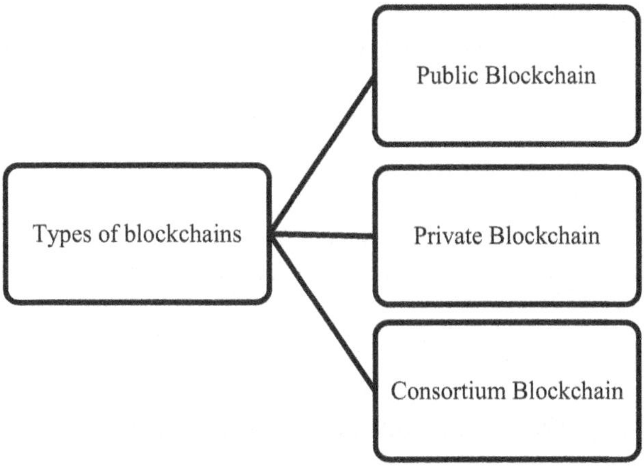

Fig. 5 Types of blockchain

glues the previous ones together by giving a reward each time a consensus has been reached and a new block is added to the chain and punishing bad actors that try to manipulate the system by taking away their token. This system works on a psychological behaviour; it turns the rules from something you need to follow into something you want to follow.

2.4 Types of Blockchain

Blockchain networks may differ in terms of who can participate and who has access to the data. They are generally ordered into three main types: public, private and consortium [14]. We present the three forms of blockchain. These three forms are shown in Fig. 5.

Public blockchain Public blockchains allow anyone to join. Checking, verifying and participating are open to everyone. These blockchains are viewed as "totally decentralized" [14]. These blockchains do not restrict participants and validators. It ensures that the data is secure because no node has complete control over the network. Bitcoin and Ethereum are both public blockchain [15].

Private blockchain In this type, it is not every node can participate; only a single organization has authority over the network and a participant needs to be invited to be a part of it. Those participants can receive varying levels of access to the blockchain. A private blockchain isn't fully decentralized. It has a set of regulations and rules that nodes have to abide by. Ripple (XRP) and Hyperledger are two examples of private blockchains.

Consortium blockchain It is partly decentralized. The node that has the authority is chosen in advance and the data can be open or private. The option to peruse the blockchain can be available to the general population or restricted by members, or "crossbreed" [14]. Quorum and Multichain are both consortium blockchains [5].

It is to highlight that when creating a blockchain application, it is essential to evaluate the type of network that will best suit business objectives. Private networks can be tightly controlled and are preferable for compliance and regulatory reasons. However, public and unauthorized networks may allow for greater decentralization and distribution.

3 How Blockchain Enhances Cybersecurity?

3.1 Blockchain Cybersecurity Situation

Many sectors are currently studying blockchain applications, which are the subject of pilot projects. Most of these new applications need to process and stock sensitive data. According to the literature, three aspects confuse the security management of the blockchain [16]:

Immaturity and complexity of technology Due to the various available consensus algorithms, and blockchains underlying complex cryptographic protocols, security professionals have difficulty understanding data flows and potential security vulnerabilities [17]. In addition, there are several platforms and implementations of blockchain, which require evaluating an application's ability to integrate a given blockchain system.

Lack of blockchain technology standards and regulations Blockchain technology is currently uncontrolled, resulting in legal uncertainties and grey areas [18]. An interesting example of the lack of controls and laws governing blockchain networks is the hacking of decentralized autonomous organizations (DAO) [19], where a vulnerability related to smart contracts has resulted in huge losses to the network.

The widespread belief that blockchain is secure by default Blockchain technology relies on public key cryptography and primitives, such as digital signatures and hash functions, which can give a false sense of security. The fact that all cryptographic protocols have their limitations and that overall security includes not only technology but also people and processes is often overlooked by blockchain security analytics.

Blockchain security is therefore a comprehensive risk management system of a blockchain network, which uses cybersecurity infrastructures, insurance services and best practices to reduce the risk of attack and fraud.

Hackers targeting the blockchain Threat actors have found ways to exploit blockchain, companies and cryptocurrency exchanges, with devastating results.

In 2018, more than $1.5 billion in cryptocurrency was lost to social engineering attacks and crypto wallet theft. Since December 2017, around 5% of all Bitcoin in circulation has been stolen by attacks on cryptocurrency exchanges. Lastly, attacks on significant cryptocurrency exchange companies such as Mt. Gox and Bitfinex have cost the companies around $350 million and $72 million, respectively. In addition to financial damage, the threat actors exploit blockchain companies in every way they can, one of which is stealing PII, personally identifiable information of customers. Blockchain companies collect a vast amount of data from their customers for security reasons. This "Know Your Customer" policy puts the companies in the scope of hackers. If the hackers steal customer data, they leverage the stolen PII by selling the data on Dark Web forums.

Top cyberattacks on blockchain-related companies in 2021
PolyNetwork, a decentralized finance (DeFi) project, was hacked in August 2021, resulting in over $600 million loss. The company has announced that more than 99% of the stolen money was returned to the firm. The attack has become the most prominent cryptocurrency theft, surpassing an attack on Coincheck in 2018, which had over $534 million stolen. After the attack, PolyNetwork asked the threat actor to become the chief security advisor to the company. Cream Finance, another decentralized finance project, was hacked three times this year. The first hack occurred in February, costing the company around $37 million. In August 2021, another hack worth $29 million took place because of an exploit in the company's source code. The last attack resulted in $130 million stolen cryptocurrency from the company in October. The third hack on Cream Finance has become the second biggest crypto attack in 2021. Liquid, a Japanese crypto exchange platform, was hit with an attack on August 2021. The hackers have stolen an estimated worth of $97 million in digital currency. After the attack, Liquid announced that some digital wallets were compromised and suspended all deposit and withdrawal operations. bZx, another DeFi platform, was hacked this year in November. After the attack, the company posted a preliminary post-mortem report stating that the hacker has used a phishing attack to steal private keys and gain access to the victim developer's wallet. The developer was phished with an email attachment and a Microsoft Word document including malicious macros. The hacker has stolen $55 million worth of cryptocurrency from the company [20].

Security challenges in traditional systems have been a longstanding concern as the reliance on digital infrastructure continues to grow. Traditional systems, which often use centralized architectures, face several vulnerabilities and limitations that make them susceptible to various cyber threats and cyberattacks. We will cite some of the significant security challenges in traditional systems: centralized points of failure, data breaches and hacks, lack of transparency, limited data integrity, single point of control, denial-of-service (DoS) attacks, inadequate authentication and authorization, legacy systems and outdated security measures, costly security upgrades and regulatory compliance concerns. For more details, see [2, 3].

Addressing these security challenges in traditional systems has become increasingly complex as cyber threats continue to evolve. In light of these limitations, blockchain technology has emerged as a promising alternative that can address many of the security concerns associated with centralized systems. Its decentralized, immutable and transparent nature makes it a compelling solution for enhancing cybersecurity across various domains.

3.2 Ways Blockchain Enhances Cybersecurity

Overall, blockchain's inherent features, such as immutability, decentralization, consensus mechanisms and smart contracts, contribute to its potential as a robust cybersecurity solution. By leveraging these properties, blockchain can significantly enhance the security and trustworthiness of digital systems, protecting sensitive data and mitigating the risks posed by various cyber threats. Blockchain technology offers several unique features that enhance cybersecurity and address the security challenges present in traditional centralized systems. Below are the ways blockchain enhances cybersecurity:

Immutable Ledger and Data Integrity Blockchain's distributed ledger maintains an immutable record of all transactions and data. Once data is recorded on the blockchain, it cannot be altered or deleted without consensus from the majority of the network. This immutability ensures data integrity, making it extremely difficult for malicious actors to tamper with records or manipulate information.

Decentralization and Resistance to Attacks Traditional systems are vulnerable to attacks targeting central points of control. In contrast, blockchain operates on a decentralized network of nodes, each holding a copy of the entire ledger. There is no single point of failure, making it resistant to attacks attempting to compromise a central authority.

Consensus Mechanisms and Byzantine Fault Tolerance Blockchain networks use consensus mechanisms, such as Proof of Work (PoW) or Proof of Stake (PoS), to validate and agree on the state of the ledger. These mechanisms ensure that all participants in the network reach a consensus on the validity of transactions. Byzantine Fault Tolerance (BFT) further strengthens the system by allowing it to function correctly even if some nodes act maliciously or fail.

Smart Contracts for Secure Automated Processes Smart contracts are self-executing contracts with predefined rules and conditions. They automate processes and execute actions automatically when specific conditions are met. Smart contracts increase security by eliminating the need for intermediaries and reducing the potential for human errors or manipulation in contract execution.

Secure Digital Identity Management Blockchain can be used for secure digital identity management, offering individuals more control over their identities and personal data. Self-sovereign identity solutions on the blockchain allow users to have ownership of their identity information, reducing the risk of identity theft and data breaches.

Decentralized Access Control Blockchain can implement decentralized access control systems, reducing the reliance on centralized identity providers and databases. Users can control their access rights through cryptographic keys and permissions stored on the blockchain, minimizing the risk of unauthorized access.

Data Integrity and Provenance Blockchain technology can timestamp data entries and create an immutable record of data provenance. This feature is particularly valuable in industries like supply chain management, ensuring the integrity and traceability of goods from their origin to the end consumer.

Supply Chain Security Blockchain can enhance supply chain security by providing transparency and traceability of goods throughout the supply chain. It helps prevent counterfeit products, ensures product authenticity and reduces the risk of fraudulent activities in the supply chain.

Internet of Things (IoT) Security Blockchain can secure IoT devices by providing decentralized identity and authentication for each device. It can also help prevent distributed denial-of-service (DDoS) attacks by distributing traffic across the network, making it harder for attackers to overwhelm a single point.

Decentralized Threat Intelligence Blockchain enables collaborative security measures, allowing different entities to share threat intelligence and detect emerging threats in real time. This collective approach to cybersecurity enhances the overall resilience of the network against attacks.

4 Cybersecurity Applications of Blockchain

4.1 Attack Methodology

We summarize the attack methodology steps discussed in Fig. 6.

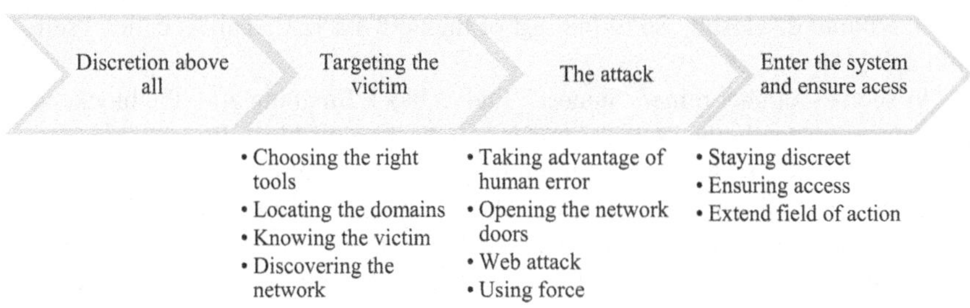

Fig. 6 Attack methodology

4.2 *Typical Blockchain Attack*

Hackers and impostors threaten blockchains in four main ways: phishing, routing, Sybil-type attacks and 51% [21, 22].

Phishing Attacks Phishing is a scam attempt to obtain user identification information. Fraudsters send wallet key owners emails designed to make them look like they came from a legitimate source. Emails ask users for their identification information using false hyperlinks. Having access to user credentials and other sensitive information can result in losses to the user and the blockchain network.

Routing attacks Blockchains are based on large, real-time data transfers. Hackers can intercept data when it is transferred to Internet service providers. In a routing attack, blockchain participants usually can't see the threat, so everything seems normal. However, behind the scenes, fraudsters have extracted confidential data or currency.

Sybil attacks In a Sybil attack, hackers create and use many fake network identities to flood the network and bring down the system. Sybil refers to a famous book character with multiple personality disorders.

51% attacks Mining requires a lot of computing power, especially for large-scale public blockchains. But if a miner, or a group of miners, could gather enough resources, it could reach more than 50% of the mining power of a blockchain network. Having more than 50% of the power means having control over the blockchain and the ability to manipulate it.

4.3 *An Example of an Attack*

One of the first sources of risk concerns is the mining procedure. When the blockchain runs under "Proof of Work", a validated block offers the miner who solved the problem a reward in the form of Bitcoin (12.5 Bitcoins per confirmed block) for the work provided. Drifts may then appear. If a miner discovers the solution first, then he has a Be block, which he should communicate to the other participants. But it can keep this block secret and work on the validation of the next block without disclosing Be to the rest of the network [23]. This is called "selfish mining" [24].

As soon as another miner, "honest", finds a block Bh going after the blockchain, then almost instantly the "selfish" miner will disclose his block Be.

The network thus finds itself in the presence of two blocks validated almost at the same time and temporarily kept on the blockchain. Some nodes in the network will be aware of the Be block and others will be aware of the Bh block. New blocks will then be added to Be's and Bh's suite. Two different chains are formed and maintained until a longer chain is identified. It is then this chain that will be kept

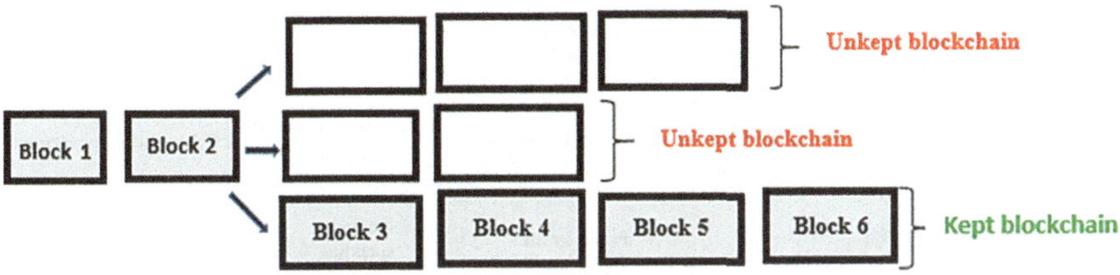

Fig. 7 Illustration of a bifurcation [24]

in the blockchain, while the other one will be deleted. The simultaneous creation of two blocks causes what is called a "bifurcation" (Fig. 7).

If A, for example, exchanges a Bitcoin with B and that transaction is recorded in Be, and, at the same time, A exchanges a Bitcoin with C in Bh, then one is in a case of double spending [25] and, finally, one of the users (B or C) will be stolen.

4.4 Blockchain Attack Classifications

General blockchain attacks Blockchain technology relies on public key cryptography and primitives such as digital signatures and hash functions, which can give a false sense of security. The fact that all cryptographic protocols have their limitations and that overall security includes not only technology but also people and processes is often overlooked by Blockchain security analytics.

To address the security provided by blockchains, it is tempting to reduce blockchain to its simplest expression. We then obtain a concept of simplicity, and even of technical poverty, extreme: it is a chained list where each new block points to the previous block. If such a construction is constantly used in computer science to store data, it does not guarantee the security of the data stored in a malicious environment and it is even more difficult to share and update such a distributed data structure.

We find in the literature different taxonomies of attacks and solutions related to blockchains. For that, we present one of them in Fig. 8. The reader can find other attacks taxonomies in [26–34].

How to protect a blockchain application? Prevention controls are used to strengthen the ability to defend significant assets against known and emerging threats. Since the pillar of the blockchain is cryptography, it seems obvious to use this technique to implement prevention controls. The key is to make sure that additional checks are done to perfect those provided by default and that these checks are done on all participating nodes. We can classify the main control and prevention areas, which need to be implemented across data, application and infrastructure layers when designing blockchain-based applications, into three broad classes:

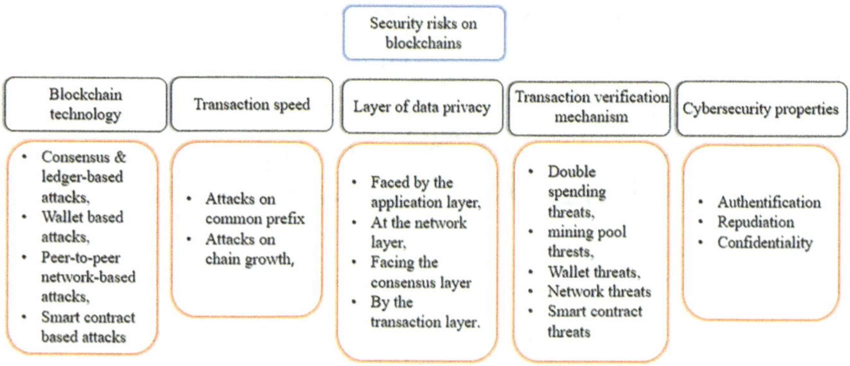

Fig. 8 Taxonomy of Blockchain-related attacks

Data Protection. Blockchain technology was designed using cryptographic primitives such as hash functions for data integrity (immutability) and digital signatures for authenticity and non-repudiation purposes. While the integrity and authenticity of the data are present from the design stage, the blockchain does not guarantee the confidentiality of the data. Digital signatures exploit public key infrastructures (PKI) which can also be used to protect data on the chain, that is to say, data stored on the blockchain itself via encryption. Other cryptographic techniques can be used to reduce or eliminate dependence on single nodes. This includes ensuring that multiple nodes collectively decrypt using shared keys or collectively sign critical data using multiple signature processes. Finally, data can be better protected through the application of data minimization, by keeping sensitive data out of the chain and allowing only non-critical data to be present on the chain.

Exploiting the existing PKI to ensure data confidentiality is tempting, but this creates a risk of availability due to the high dependence on PKI for various uses, including authentication, authorizations and data protection.

Application Protection. One of the major challenges in protecting blockchain-based applications is to train security engineers so that they understand blockchain technology, its properties and how these affect the overall safety of the systems that are developed from this technology. In addition, blockchain-based concepts, such as smart contracts, can have complex codes. The need for secure development policies and processes to ensure the use of pre-approved and tested software interfaces and libraries, regular code checks and the addition of patches is exacerbated because smart contracts are fully automated. In addition, these contracts generally receive inputs from external blockchain data, such as exchange rates or smart sensor measurements. Therefore, proper input validation and data integrity checks must be performed to protect the functionality and integrity of the systems.

Infrastructure Protection. Since blockchain technology is developed using traditional components, all typical infrastructure attack vectors, such as malware and hacking, can reach blockchain-based applications. Traditional infrastructure controls such as vulnerability analyses and patch management should therefore be performed on all nodes. It may be necessary to use virtual private network (VPN) gateways to establish connectivity between nodes and allow geographically dispersed nodes to communicate securely.

Cybersecurity applications of blockchain leverage the technology's unique features to enhance security and protect against various cyber threats. Below are some of the key cybersecurity applications of blockchain:

4.5 Key Cybersecurity Applications

Secure Digital Identity Management Blockchain can be used for secure digital identity management, allowing individuals to have more control over their identities and personal data. Self-sovereign identity solutions on the blockchain enable users to manage and authenticate their identity information without relying on a centralized authority. This reduces the risk of identity theft, data breaches and unauthorized access to personal information.

Decentralized Access Control Blockchain can enable decentralized access control systems. Users can control access to their data and resources through cryptographic keys and permissions stored on the blockchain. This reduces the reliance on centralized identity providers and enhances security by minimizing the risk of unauthorized access.

Data Integrity and Provenance Blockchain's immutable ledger ensures data integrity and creates an audit trail for data provenance. Timestamping and recording data on the blockchain make it challenging for malicious actors to alter or manipulate data, providing a reliable and tamper-proof record of information.

Supply Chain Security Blockchain technology can be applied to supply chain management to ensure transparency and traceability of goods. By recording the entire supply chain process on the blockchain, stakeholders can verify the authenticity and origin of products, reducing the risk of counterfeit goods and ensuring supply chain integrity.

Internet of Things (IoT) Security Blockchain enhances the security of IoT devices by providing decentralized identity and authentication. Each IoT device can have a unique identity and cryptographic key stored on the blockchain, ensuring secure communication and preventing unauthorized access to IoT networks.

Decentralized Threat Intelligence Blockchain enables collaborative threat intelligence sharing among different entities. By sharing real-time threat information on the blockchain, organizations can detect and respond to emerging cyber threats more effectively, enhancing the overall security posture of the network.

Secure Data Sharing and Collaboration Blockchain facilitates secure data sharing and collaboration between multiple parties. It allows data to be shared on a need-to-know basis, and smart contracts can enforce predefined rules for data access and usage, ensuring data privacy and preventing unauthorized sharing.

Encrypted Communication and Messaging Blockchain-based applications can offer encrypted communication and messaging services, ensuring the confidentiality and privacy of sensitive communications between users.

Cyber Incident Response and Recovery Blockchain can be used to create immutable logs of cyber incidents and responses, providing an auditable trail of actions taken during incident response. This helps in understanding the attack patterns and improving incident response strategies.

Decentralized Firewalls and Intrusion Detection Systems Blockchain technology can be utilized to create decentralized firewalls and intrusion detection systems, distributed across multiple nodes. This approach improves resilience against attacks by eliminating single points of failure and centralization.

By leveraging these cybersecurity applications, blockchain technology offers novel solutions to address the security challenges faced by traditional systems, providing enhanced protection and resilience in the digital landscape

5 Case Studies: Real-World Implementations

Case studies of real-world implementations of blockchain in cybersecurity showcase the practical applications and benefits of the technology.

5.1 Online Book Sale Dapp

In this distributed application, we are building a decentralized application that will eliminate third parties and automate e-commerce processes, which can be done using a system based on blockchain principles and smart contracts that promise easy, fast and secure transactions for all online interactions. In addition to eliminating the costly infrastructure used to manage payments, the use of smart contracts

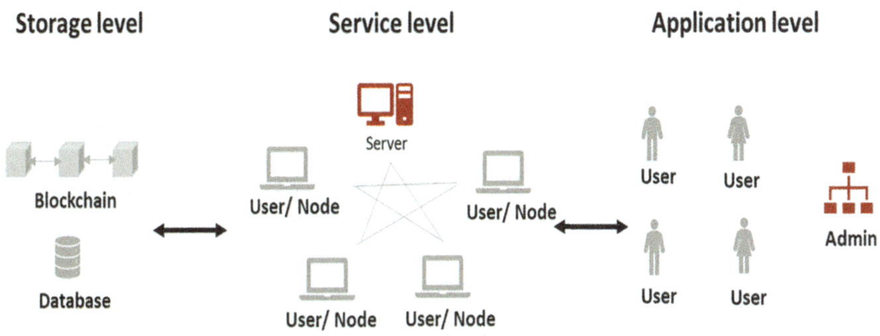

Fig. 9 System architecture

builds mutual trust between buyers and sellers and also eliminates the need to give intermediaries sensitive personal and financial information.

This system will have the same functionality as the conventional systems used, but with cheaper and faster payment processing using blockchain. So we built a system that can implement data exchange between a smart contract and customer currencies and support ETH and custom token purchases so that you can buy books using these tokens. Figures 9, 10, 11, and 12 show, respectively, the system architecture, UML diagrams, Dapp windows and deployment code of the smart contract [35].

Tools and programming languages To implement our Dapp, we used a battery of tools and languages: Next.js, Moralis, Hardhat, OpenZeppelin Contracts, MetaMask, Solidity, JavaScript and npm package.

In this Dapp, the focus was more on the smart contract and blockchains which make them susceptible to security risks. Smart contracts might contain coding flaws or vulnerabilities that hackers could use to gain access without authorization or tamper with the contract's execution. They may be the target of attacks like denial-of-service attacks or replay attacks [24] that can prevent or affect the execution of the contract. Standardization is required for the creation and use of smart contracts to reduce these risks. Smart contracts can be developed securely and reliably with clear specifications and coding standards. The adoption of standards can also encourage interoperability and make it easier to integrate smart contracts with other platforms and systems.

(a) Use case Diagram

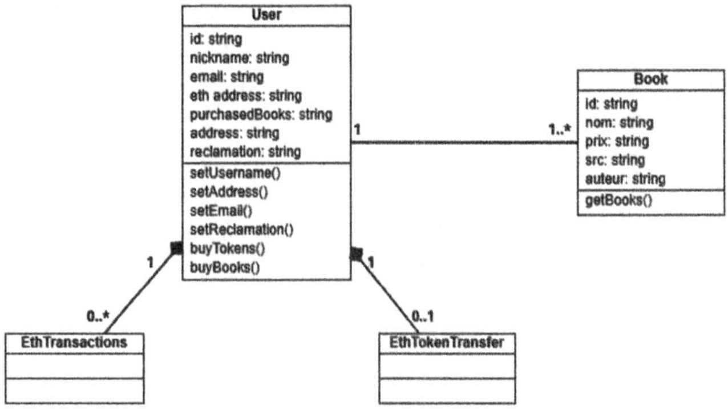

(b) Class diagram

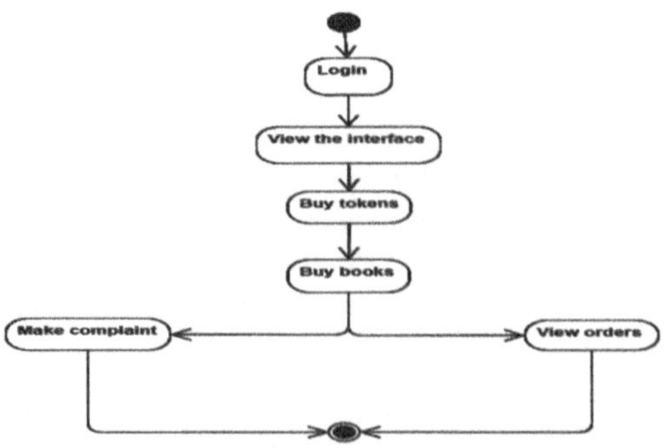

(c) Activity diagram

Fig. 10 System design with UML diagrams. (**a**) The use case diagram describes the functions and the scope of the system in addition to the interactions between the system and its two actors user and admin. (**b**) The classes and interfaces of systems and their relationships, i.e. the class user: as soon as the user is connected, all his data on the channel is instantly synchronized in the Moralis database. (**c**) The user must log into the system to have the ability to browse books, buy tokens and buy books

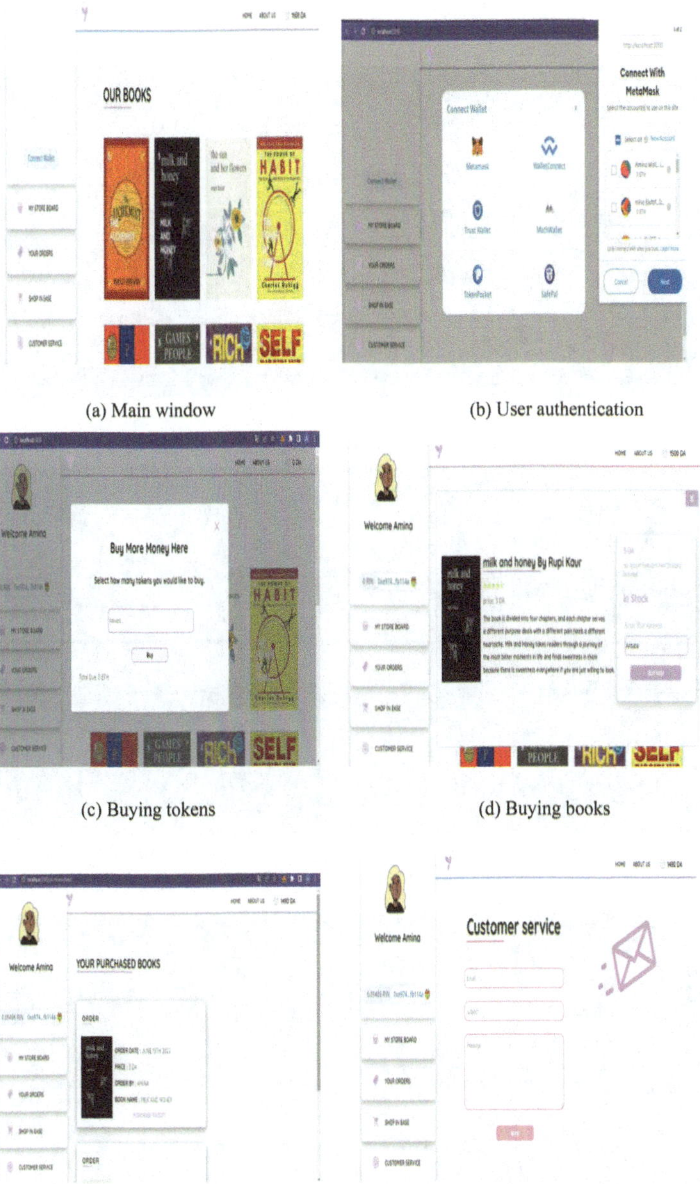

(a) Main window

(b) User authentication

(c) Buying tokens

(d) Buying books

(e) Purchased books

(f) Customer Service

Fig. 11 The system is an e-commerce Dapp and its business logic is built based on the Ethereum smart contract. It is an online bookstore where customers can buy books and pay with special tokens. (**a**) This is the main interface of the application, where it is possible to log in, start the purchase process, buy tokens or simply consult the bookstore. (**b**) User authentication, i.e. when the user logs into the Dapp using cryptocurrency wallet authentication, the address of that wallet will be automatically saved in the database Moralis with all configured data, such as token balance, transaction history or events. (**c**) Description of the purchase of special tokens by the user ether exchange. (**d**) Description of purchasing books: when there is a sufficient balance, the user can purchase tokens to pay for books. (**e**) Once the book is purchased, viewing the order and receipt of purchased books. (**f**) The customer can make a claim in case the book has not been delivered

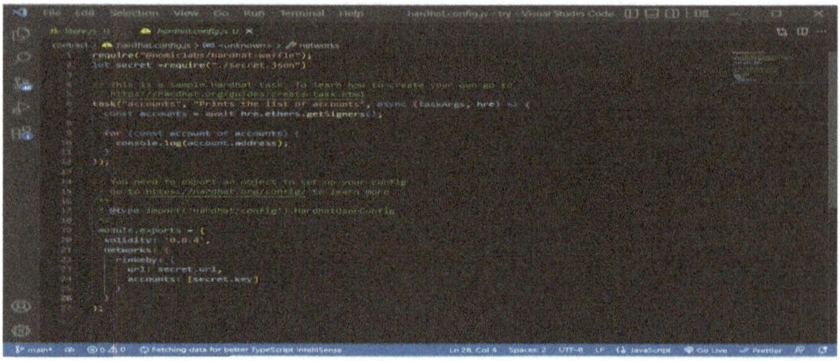

(a) Hardhat.config.js fichier

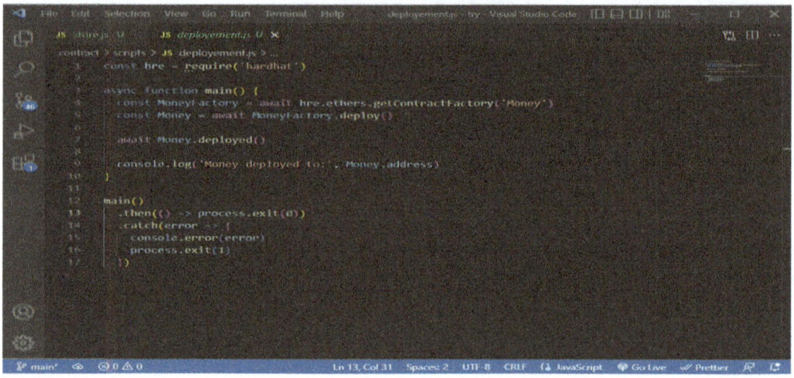

(b) deployement.js file

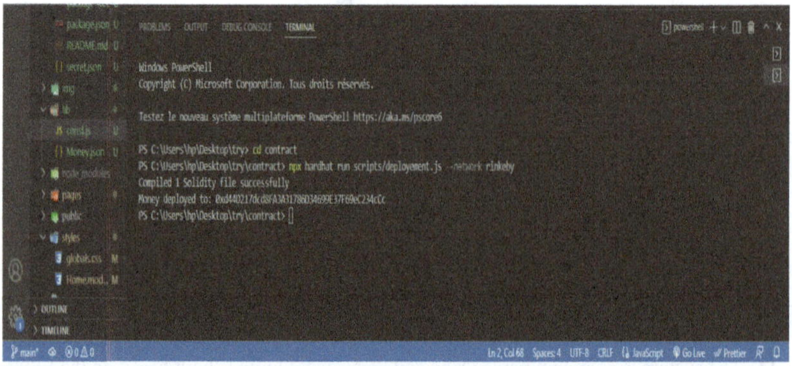

(c) Deployment of contract

Fig. 12 Main codes for smart contract deployment. (**a**) After updating, the code looks like this content. (**b**) To compile the npx hardhat compile contract: write the deployment script that will be called deployment.js and add the content above. (**c**) Calling deploy() on a ContractFactory will initiate deployment and return a *Promise* that resolves to a contract object. This is the object that has a method for each of our smart contract functions

5.2 Other Applications

Below are some examples of how blockchain has been deployed to enhance cybersecurity in different industries:

Secure Digital Identity Management – uPort uPort is a self-sovereign identity platform built on the Ethereum blockchain. It allows users to create and manage their digital identities without the need for a centralized identity provider. Users can control access to their personal information and selectively share identity attributes with trusted parties. This decentralized approach to identity management reduces the risk of identity theft and provides a more secure and privacy-conscious solution.

Decentralized Access Control – Microsoft's Identity Overlay Network (ION) Microsoft's identity overlay network is a blockchain-based solution that aims to improve decentralized identity management. ION leverages the Bitcoin blockchain to create a decentralized public key infrastructure (PKI) that enables users to manage and control their digital identities without relying on centralized identity providers. By using blockchain's immutability and consensus mechanisms, ION enhances the security and resilience of access control systems.

Data Integrity and Provenance – Walmart's Food Traceability on Hyperledger Fabric Walmart, in collaboration with IBM, implemented a food traceability solution on the hyperledger fabric blockchain. The system tracks the provenance of food products from the farm to the store shelves. By recording each step of the supply chain on the blockchain, Walmart ensures the integrity of the data and provides consumers with transparent information about the origin and quality of the products they purchase.

Supply Chain Security – Everledger's Diamond Tracking Everledger, a startup, developed a blockchain-based solution for tracking and verifying the authenticity of diamonds. By recording the unique attributes of each diamond on the blockchain, the system creates a digital "passport" for the diamond, reducing the risk of counterfeit diamonds in the supply chain and providing consumers with confidence in the authenticity of their purchases.

Internet of Things (IoT) Security – Filament's Decentralized IoT Networks Filament is a company that provides blockchain-based solutions for secure IoT networks. Their platform enables IoT devices to communicate securely with each other, leveraging blockchain's decentralized architecture and cryptographic features to prevent unauthorized access and ensure the integrity of IoT data.

Decentralized Threat Intelligence – AlienVault's Open Threat Exchange (OTX) AlienVault's Open Threat Exchange is a crowd-sourced threat intelligence platform built on blockchain technology. Security researchers and organizations can contribute and access real-time threat data on the blockchain, allowing for better collaboration and faster detection of emerging threats.

Secure Data Sharing and Collaboration – Enigma's Privacy-Preserving Solutions Enigma is a project that focuses on providing privacy-preserving solutions for secure data sharing and collaboration. Their protocol allows data to be securely shared and analysed across multiple parties without revealing the raw data to any participant, ensuring data privacy while still enabling meaningful insights.

These case studies demonstrate the diverse applications of blockchain in enhancing cybersecurity across various industries. By leveraging blockchain's inherent features, these implementations offer improved security, transparency and trust in digital systems and processes. As blockchain technology continues to evolve, more innovative use cases are expected to emerge, further strengthening cybersecurity measures in the digital age.

6 Discussions

6.1 Challenges and Limitations of Blockchain

Despite its many advantages, blockchain technology also faces several challenges and limitations that need to be addressed for its wider adoption and successful implementation. Some of the key challenges and limitations of blockchain are as follows: scalability, transaction speed, energy consumption, regulatory and legal concerns, interoperability, security concerns, governance and consensus, storage requirements, user experience, limited smart contract capabilities and environmental impact. Addressing these challenges and limitations is essential for the continued growth and adoption of blockchain technology. Efforts are being made by the blockchain community and researchers to develop innovative solutions and overcome these obstacles, making blockchain more scalable, efficient, secure and user-friendly for a wide range of applications [36, 37].

6.2 Future Perspectives of Blockchain

The future perspectives of blockchain hold immense potential for transformative changes across various industries and domains. As technology continues to evolve, several trends and developments are expected to shape its trajectory in the coming years. Here are some key future perspectives on blockchain:

Scalability and Performance Improvements Scaling solutions such as sharding, sidechains and layer-two protocols (e.g. Lightning Network for Bitcoin) are being developed to address blockchain's scalability limitations. These improvements will enable blockchain networks to process a higher number of transactions per second, making them more suitable for enterprise-level applications.

Interoperability and Cross-Chain Communication Efforts to achieve interoperability between different blockchain networks will enable seamless communication and data exchange between them. Interoperability solutions will break down silos and facilitate cross-chain asset transfers, fostering greater collaboration and efficiency.

Evolving Consensus Mechanisms While Proof of Work (PoW) and Proof of Stake (PoS) are the most well-known consensus mechanisms, ongoing research and development are exploring novel consensus algorithms. This includes mechanisms like Proof of Authority (PoA), Delegated Proof of Stake (DPoS) and Practical Byzantine Fault Tolerance (PBFT), which aim to improve security, scalability and energy efficiency.

Integration with Emerging Technologies Blockchain technology is likely to integrate with other emerging technologies such as artificial intelligence (AI), the Internet of things (IoT) and quantum computing. The combination of blockchain with AI can enhance data analysis and pattern recognition while integrating with IoT can enable secure device communication and data sharing.

Decentralized Finance (DeFi) Expansion DeFi applications have gained significant traction, offering various financial services without intermediaries. The future will likely see the expansion of DeFi into traditional finance, leading to more efficient and inclusive financial systems.

Central Bank Digital Currencies (CBDCs) Several countries are exploring the development and issuance of their central bank digital currencies. CBDCs, built on blockchain or distributed ledger technology, have the potential to revolutionize payment systems and improve financial inclusion.

Privacy Enhancements Advancements in privacy-focused blockchains, such as zero-knowledge proofs and secure multiparty computation, will address concerns about data privacy and confidentiality in public blockchain networks.

Regulatory Clarity and Frameworks Governments and regulatory bodies are gradually providing more clarity on how blockchain and cryptocurrencies are treated under existing legal frameworks. Regulatory certainty will likely foster greater institutional involvement and investment in the blockchain space.

Sustainability and Green Blockchain Initiatives Efforts to address the environmental impact of blockchain, particularly energy consumption in PoW-based networks, will lead to the emergence of more sustainable and eco-friendly blockchain solutions.

Decentralized Internet and Web 3.0 Blockchain technology's decentralized nature aligns with the vision of a more open and censorship-resistant Internet. Web 3.0, powered by blockchain and decentralized protocols, will provide users with greater control over their data and online interactions.

Supply Chain Transformation Blockchain's role in supply chain management is expected to expand further, enabling end-to-end traceability, transparency and efficiency in global supply chains.

Overall, the future of blockchain is bright, with ongoing research and development driving innovations and breakthroughs. As scalability, privacy and regulatory challenges are addressed, blockchain is poised to play an increasingly crucial role in reshaping industries and empowering individuals in the digital age.

7 Conclusion

Throughout this chapter, we explored the intersection of blockchain technology and cybersecurity, highlighting the unique features that make blockchain a robust solution for enhancing security in the digital world.

The future outlook for blockchain cybersecurity is promising, with several trends and developments shaping the path forward: advancements in scalability, interoperability and collaboration, enhanced privacy solutions, institutional adoption and so on. Blockchain technology holds immense potential to revolutionize cybersecurity by providing secure, transparent and decentralized solutions to the challenges faced by traditional systems. As the technology continues to evolve and overcome its limitations, we can expect to see blockchain play an increasingly vital role in safeguarding digital assets, ensuring data integrity and enhancing trust and security in our interconnected world. The future of blockchain cybersecurity is filled with possibilities, making it an exciting frontier for innovation and advancement in the digital age.

References

1. Chin, K.: The Role of Cybersecurity in Blockchain Technology, upguard. (2023)
2. Nakamoto, S.: Bitcoin: A Peer-to-Peer Electronic Cash System, white paper. (2008)
3. Priyadarshin, I.: Introduction to Blockchain Technology. Wiley (2019). https://doi.org/10.1002/9781119488330.ch6
4. Yang, D., Long, C., Xu, H., Peng, S.: A review on scalability of blockchain. In: ICBCT. The 2nd International Conference on Blockchain Technology, pp. 1–6, Hawaii, USA (2020)
5. Lin, L.C., Liao, L.C.: A survey of blockchain security issues and challenges. Int. J. Netw. Secur. **19**, 653–659 (2017)
6. Bouraga, S.: A taxonomy of blockchain consensus protocols: a survey and classification framework. Expert Syst. Appl. **168**, 114384 (2021)
7. Pahlajani, S., Kshirsagar, A., Pachghare, V.: Survey on private blockchain consensus algorithms. In: 2019 1st International Conference on Innovations in Information and Communication Technology, pp. 1–6, Chennai, India (2019)
8. Zhang, S., Lee, J.H.: Analysis of the main consensus protocols of blockchain. ICT Express. **6**, 93–97 (2020)

9. Ahmed-Rengers, M., Kostiainen, K.: Don't mine, wait in line: fair and efficient blockchain consensus with robust round robin. arXiv preprint arXiv:1804.07391. (2018)
10. Liu, J., Li, W., Karame, G.O., Asokan, N.: Scalable byzantine consensus via hardware-assisted secret sharing. IEEE Trans. Comput. **68**, 139–151 (2018)
11. Bravo-Marquez, F., Reeves, S., Ugartezx, M.: Proof-of-learning: a blockchain consensus mechanism based on machine learning competitions. In: Proc of the 2019 IEEE International Conference on Decentralized Applications and Infrastructures, pp. 4–9, Newark, CA, USA (2019)
12. Bamakan, S.M.H., Motavali, A., Bondarti, A.B.: A survey of blockchain consensus algorithms performance evaluation criteria. Expert Syst. Appl. **154**, 113385 (2020)
13. Bach, L.M., Mihaljevic, B., Zagar, M.: Comparative analysis of blockchain consensus algorithms. In: 41st International Convention on Information and Communication Technology, Electronics and Microelectronics (MIPRO), pp. 1545–1550, Opatija, Croatia (2018)
14. Muda, I., Madem, S., Hasan, S., Ahmod, S., Kayande, R.A., Chakraborty, N.: A survey on applications and security issues of blockchain technology in business sectors. In: 2022 2nd International Conference on Technological Advancements in Computational Sciences (ICTACS), pp. 663–668 (2022)
15. Tang, H., Shi, Y., Dong, P.: Public blockchain evaluation using entropy and TOPSIS. Expert Syst. Appl. **117**, 204–210 (2019)
16. Saberi, S., Kouhizadeh, M., Sarkis, J., Shen, L.: Blockchain technology and its relationships to sustainable supply chain management. Int. J. Prod. Res. **57**(7), 2117–2135 (2019)
17. Zamani, E., He, Y., Phillips, M.: On the security risks of the blockchain. J. Comput. Inf. Syst. **60**(6), 495–506 (2020)
18. Egger, P., Karaklajic, D., Julisch, K., Widmer, F.: Blockchain Security: Protecting the Distributed Ledger. Deloitte AG (2019)
19. Vessenes, P.: More Ethereum Attacks: Race-To-Empty is the Real Deal, On Ethereum, Smart Contracts, Security, Solidity, 9 June 2016
20. Principales cyberattaques contre les échanges de crypto-monnaie et les sociétés de blockchain en 2021 – SOCRadar® Cyber Intelligence Inc.
21. Hasanova, H., Baek, U.J., Shin, M.G., Cho, K., Kim, M.S.: A survey on blockchain cybersecurity vulnerabilities and possible countermeasures. Int. J. Netw. Manag. **29**, e2060 (2019)
22. Bancal, D., Dumas, D., Puche, D., Ebel, F., Vicogne, F., Fortunato, G., Hennecart, J., Schalkwijk, L., Agé, M., Rault, R., Crocfer, R., Lasson, S.: Sécurité informatique Ethical Hacking: Apprendre l'attaque pour mieux se défendre. Editions ENI, France (2015)
23. Eyal, I., Sirer, E.G.: Majority is not enough: Bitcoin mining is vulnerable. Commun. ACM. **61**(7), 95–102 (2018)
24. Göbel, J., Keeler, H.P., Krzesinski, A.E., Taylor, P.G.: Bitcoin blockchain dynamics: the selfish-mine strategy in the presence of propagation delay. Perform. Eval. **104**, 23–41 (2016)
25. Decker, C., Wattenhofer, R.: Information propagation in the bitcoin network. In: Conference IEEE P2P 2013 Proceedings
26. Aboujaoude, J., Saade, R.: Business applications of blockchain technology: a systematic review. IEEE-Access, p. 10 (2019)
27. Zheng, Z., Xie, S., Dai, H., Chen, X., Wang, H.: An overview of blockchain technology: architecture, consensus, and future trends. In: 2017 IEEE International Congress on Big Data, BigData Congress, pp. 557–564. IEEE (2017)
28. Anand, A., Raj, A., Kohli, R., Bibhu, V.: Proposed symmetric key cryptography algorithm for data security. In: 2016 International Conference on Innovation and Challenges in Cyber Security (ICICCS-INBUSH), pp. 159–162, Noida (2016). https://doi.org/10.1109/ICICCS.2016.7542294
29. Kiayias, A., Panagiotakos, G.: Speed-security tradeoffs in blockchain protocols. Cryptology ePrint Archive (2015)
30. Zhang, Q., Lv, H., Ma, J., Li, J., Zhang, J.: Overview of Blockchain Data Privacy Protection, BIOTC 2021, Ho Chi Minh City, Vietnam, 8–10 July 2021

31. Nayak, K., Kumar, S., Miller, A., et al.: Stubborn mining: generalizing selfish mining and combining with an eclipse attack. In: IEEE European Symposium on Security and Privacy, pp. 305–320. IEEE (2016). https://doi.org/10.1109/EuroSP.2016.32
32. Zhang, H., Duan, H., Jianping, W.U.: IP clustering based reputation system for anti-spam. J. Tsinghua Univ. **50**(10), 1723–1727 (2010)
33. Rot, A., Blaicke, B.: Blockchain's future role in cybersecurity. Analysis of defensive and offensive potential leveraging blockchain-based platforms. In: 2019 9th International Conference on Advanced Computer Information Technologies (ACIT). IEEE, Piscataway (2019)
34. Gimenez-Aguilar, M., de Fuentes, J.M., Gonzalez-Manzano, L., et al.: Achieving cybersecurity in blockchain-based systems: a survey. Futur. Gener. Comput. Syst. **124**, 91–118 (2021)
35. Echchaoui, H., Boudour, R.: Online book selling Dapp based on smart contracts. In: Artificial Intelligence Doctoral Symposium (2023). https://doi.org/10.1007/978-981-99-4484-2_14
36. Zheng, Z., Xie, S., Dai, H.N., Chen, X., Wang, H.: Blockchain challenges and opportunities: a survey. Int. J. Web Grid Serv. **14**, 352–375 (2018)
37. Sanka, A.I., Cheung, R.C.C.: A systematic review of blockchain scalability: issues, solutions, analysis and future research. J. Netw. Comput. Appl. **195**, 103232 (2021)

Part II
Blockchain for Enhanced E-Government Services

Blockchain-Integrated Secure Framework for Enhanced E-Government Services

Sajedul Talukder, Md Jahangir Alam, Ismail Hossain, and Sai Puppala

Abstract As the digital transformation of public services intensifies, e-government systems face significant challenges in identity verification and data security. This chapter proposes a blockchain-integrated identity system as a robust solution to these issues, thereby enhancing the efficiency and security of e-government services. By leveraging the decentralized, transparent, and immutable characteristics of blockchain technology, this system presents an innovative approach to identity management that reduces the risk of identity theft and fraud. The chapter explores the conceptual framework of the blockchain-integrated identity system, discusses its potential benefits and implementation challenges, and provides insights from real-world case studies. The findings underscore the considerable potential of blockchain technology in revolutionizing e-government services by offering more reliable, secure, and user-friendly identity verification processes.

Keywords E-government · Zero-knowledge proof · Blockchain · Identity solution

1 Introduction

The integration of digital technologies into public services, commonly referred to as e-government, has revolutionized the delivery of government services worldwide. It has not only streamlined operations but also increased accessibility, making it easier for citizens to interact with their governments. While these developments are highly beneficial, they also present significant challenges, particularly in the realm of identity verification and data security. This chapter introduces a blockchain-integrated identity system as a solution to these challenges, thereby enhancing

S. Talukder (✉) · M. J. Alam · I. Hossain · S. Puppala
School of Computing, Southern Illinois University, Carbondale, IL, USA
e-mail: sajedul.talukder@siu.edu; mdjahangir.alam@siu.edu; ismail.hossain@siu.edu; saimaniteja.puppala@siu.edu

N. El Madhoun et al. (eds.), *Building Cybersecurity Applications with Blockchain and Smart Contracts*, Signals and Communication Technology,
https://doi.org/10.1007/978-3-031-50733-5_2

e-government services. The world is witnessing an exponential increase in digital identities, with individuals often possessing multiple digital personas across various platforms. The management and validation of these identities have become complex tasks for governments, frequently leading to issues of identity theft and fraud. Thus, the need for a more secure, efficient, and reliable identity verification system has never been more crucial.

Blockchain technology, with its core features of decentralization, transparency, and immutability, offers a robust solution to these challenges. Blockchain can create a highly secure environment where each user's identity information is stored in a decentralized ledger, reducing the risk of a single point of failure. Its transparency ensures the auditability of transactions, allowing citizens to control and verify how their personal information is used. At the same time, its immutable nature ensures that once an identity is verified and recorded on the blockchain, it cannot be altered or deleted, thus mitigating the risk of identity fraud. In this context, the chapter proposes a blockchain-integrated identity system that leverages these features of blockchain technology to enhance e-government services. By adopting such a system, governments can significantly improve the security and efficiency of their identity verification processes, leading to more reliable, secure, and user-friendly e-services. This system could revolutionize how citizens interact with their government, providing a seamless experience while ensuring their data's utmost privacy and security.

This chapter delves into the intricacies of such a blockchain-integrated identity system, discussing its components, potential benefits, and the challenges faced during its implementation. Our comprehensive exposition outlines an e-government framework attuned to contemporary challenges and future needs. While the framework represents a significant leap in e-government operations, future endeavors must focus on quantum-resistant cryptographic algorithms and harnessing the Internet of Things (IoT) for broader, interconnected public service ecosystems. Ultimately, the aim is to explore the full potential of blockchain technology in enhancing the security and efficiency of e-government services through a more robust identity verification system. In the following sections, we delve into the specifics of blockchain technology and its applications in identity management, present our proposed model of a blockchain-integrated identity system, discuss its potential benefits and challenges, and explore its practical implications through real-world case studies.

The structure of this chapter unfolds as follows. Initially, we provide an in-depth background context setting the stage for our discourse. Following that, we delve into the pertinent literature, laying the groundwork for the insights presented in this chapter. Subsequently, we unveil our proposed e-government framework, which is central to the discussions herein. The succeeding section is dedicated to elucidating the user registration and authentication mechanisms within our blockchain-based system. We then explore the multifaceted applications of integrating blockchain technology within various e-government services. The chapter culminates with a conclusive synthesis of our discussions, supplemented with insights into potential future avenues.

2 Background

E-government refers to the use of digital technologies to improve public service delivery and citizen engagement with the government. It encompasses a wide range of applications, such as online portals for paying taxes, applying for permits, and accessing public records. The benefits of e-government include increased efficiency, transparency, and accessibility of public services, as well as better citizen participation and accountability. Blockchain technology, on the other hand, is a distributed ledger system that enables secure and transparent record-keeping without the need for intermediaries. It uses cryptographic techniques to ensure the integrity and authenticity of transactions, and it is resistant to tampering and fraud. Blockchain technology has gained significant attention in recent years due to its potential to disrupt various industries, including finance, healthcare, and supply chain management. In the context of e-government, blockchain technology offers several advantages over traditional record-keeping systems. First, it can provide a secure and transparent way of storing and sharing sensitive data, such as personal identification, land registry, and voting records. Second, it can enable decentralized and tamper-proof decision-making processes, such as smart contracts and digital signatures. Third, it can enhance the trust and accountability of government institutions by providing immutable and auditable records of their activities.

The opportunities presented by blockchain technology in e-government include increased transparency, efficiency, and citizen participation. The use of blockchain technology can increase transparency in government operations, reducing corruption and increasing public trust. The decentralized nature of blockchain technology can streamline government operations, reducing bureaucracy and increasing efficiency. By creating blockchain-based systems that allow for secure and transparent participation, citizens can have a more active role in the decision-making process. However, the adoption of blockchain technology in e-government also presents various challenges and opportunities. Some of the challenges include technical challenges, regulatory challenges, and cultural challenges. Implementing blockchain technology requires a significant investment in infrastructure and technical expertise. The regulatory environment for blockchain technology is still evolving, and government agencies must navigate a complex web of regulations. Moreover, the adoption of blockchain technology requires a cultural shift in government agencies and the broader society. In this section, we describe the basic concepts and technologies of blockchain that we leverage in this chapter. More precisely, we discuss the background on blockchain, smart contract, consensus algorithms, etc.

2.1 Blockchain

A blockchain is a decentralized ledger technology that enables secure and transparent record-keeping of transactions. The concept of blockchain was first introduced

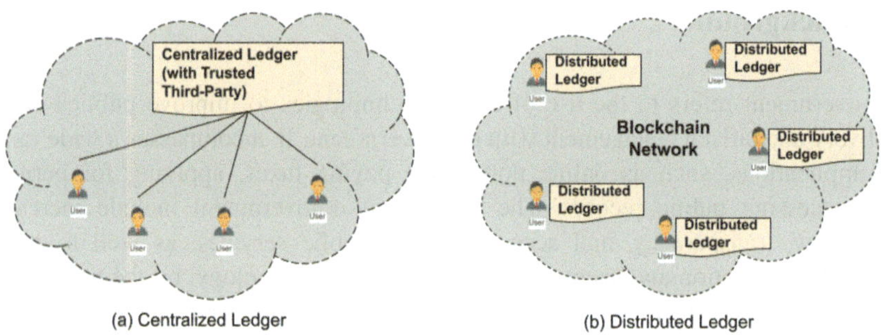

Fig. 1 (**a**) Traditional (centralized) ledger consists of a single central ledger for all users. (**b**) Distributed ledger consists of separate ledgers for each users [2]

by Satoshi Nakamoto in 2008 in the cryptocurrency project bitcoin [15]. Unlike traditional ledger system that are centralized, a blockchain is distributed that is spread across a network of nodes. Figure 1 depicts the traditional (centralized) vs. distributed ledger.

In a blockchain each node stores the same copy of the ledger. One of the key features of a blockchain is its immutability. Once a transaction is recorded on the blockchain, it cannot be altered. Because any changes to the ledger are validated and recorded through a consensus mechanism. A consensus mechanism is a process by which multiple nodes in a blockchain network agree on the validity of each transaction to the blockchain.

Consensus mechanisms are critical to the security and integrity of the blockchain network, and there are different consensus mechanism. Different algorithms have different trade-offs between security and integrity. Here we use algorithm and mechanism interchangeably. Some examples of consensus algorithms include proof of work (PoW) [8], proof of stake (PoS) [20], and delegated proof of stake (DPoS) [11]. These mechanisms ensure that all nodes in the network have a shared view of the state of the blockchain and that changes are agreed upon by the majority of nodes. This makes blockchain secure and transparent. This also ensures that the data stored on the blockchain is tamper-proof. In general in a blockchain, transactions occur following the blockchain process consisting of several steps depicted in Fig. 2.

In a blockchain network, nodes communicate by sending and receiving transactions. A transaction is requested by a node and is authenticated by the other nodes in the blockchain network. A transaction is a data package containing information about the sender, receiver, and content the sender wants to send to the receiver. Nodes can send transactions to other nodes or smart contracts (SCs). An SC is simply a program stored on a blockchain network that runs when predetermined conditions are met. It can be triggered by transactions received from nodes or other SCs. The process of transaction is completed by creating a block in the chain; block gets validated by other nodes. The node which validates the block gets reward as a result of a proof-of-work system. Figure 3 shows a typical blockchain with three blocks.

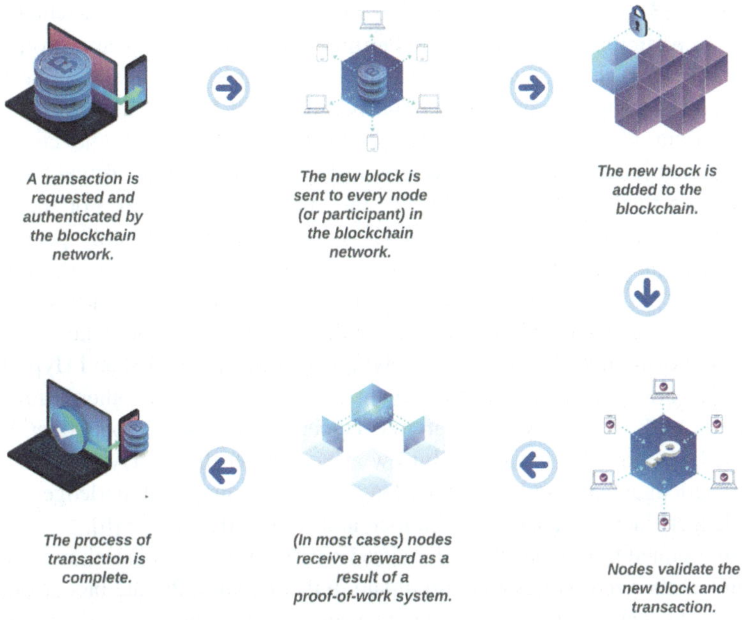

Fig. 2 Blockchain process

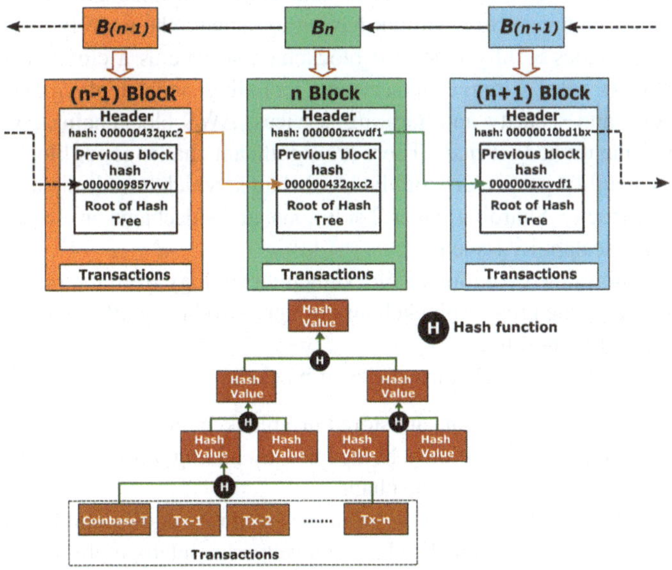

Fig. 3 Blockchain architecture with three blocks

Based on the authorization to participate in the blockchain network, there are two types of blockchains, permissioned and permissionless. Bitcoin [15] and

Ethereum [28] are two major permissionless blockchain platforms which allow anyone to participate in the network. Permissionless blockchains are also referred to as public blockchains. On the other hand, permissioned blockchains are restricted to a specific group of participants. The permissioned blockchain is divided into a consortium blockchain and a private blockchain. Generally, the consortium blockchain is jointly managed by several institutions and can determine the degree of openness to the public according to the application scenario. For the private blockchain, data are transparent to only system members, and only people with permission can join. The private blockchain is not fully decentralized, and some parties have more power than other nodes. Commonly, some parties have the right to allow nodes to join a network, or they can be validators who create data blocks and commit to the blockchain. Hyperledger Sawtooth [16] and Hyperledger Fabric [1] are typical private blockchain systems. There are other frameworks under the umbrella of hyperledger like Hyperledger Besu, Hyperledger Fabric, Hyperledger Indy, Hyperledger Burrow, Hyperledger Iroha, Hyperledger Grid, and Hyperledger Labs and five Hyperledger tools, including Hyperledger Avalon, Hyperledger Cactus, Heperledger Caliper, and Hyperledger Cello [6].

Permissioned blockchains are often used by enterprises or consortiums to ensure that only authorized parties can participate in the network. Private blockchains are also used by enterprises or government agencies to ensure privacy and control. There are some cloud services available now for blockchain. Blockchain as a service (BaaS) is a cloud-based service that allows organizations to deploy and manage blockchain networks without having to manage the underlying infrastructure. AWS provides a fully managed blockchain service as a cloud service [22]. BaaS providers offer a range of services, including node deployment, network management, and security and auditing services. AWS blockchain service offers all the abovementioned services. There are significant challenge for blockchain networks in scalability, particularly public blockchains, which have limited transaction processing capacity. Sharding, layer 2 solutions, and sidechains are some solutions to address the scalability issue [12].

Figure 4 shows the evolution of blockchain technology from the proposal phase in year of 1982 to the present blockchain through introducing different versions and upgrading at different time.

We can summarize the characteristics of blockchain as below points [7]:

- **Immutability:** Once the data are stored in a blockchain, it is infeasible to change or remove them. Data can only be updated by adding a new block, but the old version is still maintained in the chain.
- **Interoperability:** Interoperability refers to the ability of different blockchains to communicate and exchange data between networks. Interoperability is important for enabling cross-chain transactions and creating a seamless experience for users.
- **Decentralized:** No entity entirely governs the blockchain network. Data are recorded to the chain by miners or validators if the transaction satisfies the network's policy and is not tampered with by a malicious node. No one in the

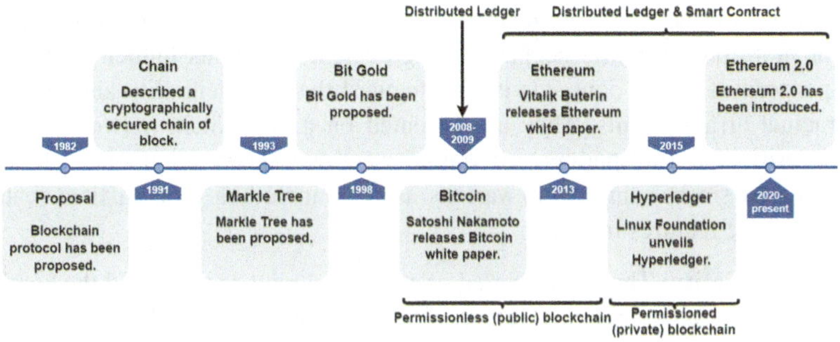

Fig. 4 Evolution of blockchain

network can personally add a block to the chain or deny a valid block to be added.

- **Enhanced Security:** On top of decentralization cryptography is another layer of protection for the blockchain network. Every data on the blockchain is hashed cryptographically, and this works as a firewall for attacks. All the blocks in the ledger come with a unique hash of their own and contain the hash of the previous block. So, changing or trying to tamper with the data will mean changing all the hash IDs. And that's kind of impossible.
- **Distributed Ledger:** This is another major feature of blockchain. All the nodes on the network contain the same view of the ledger, and each node does the necessary computation itself. This distributed computation power across the nodes ensures better outcome.
- **Consensus:** A block is considered valid if and only if more than half the number of nodes in the network agree with its validity. The chain of these valid blocks is the main chain. Blocks that are not in the main chain are orphan blocks. Orphan blocks are correctly minted, and the data in them are corrected.

2.2 *Smart Contract*

Smart contracts represent a significant advancement in blockchain technology. They were originally proposed in the 1990s as a digital transaction protocol to execute the terms of an agreement [19]. Essentially, smart contracts are bundles of code that capture and replicate the terms of real-world contracts in a digital format. A contract is essentially a legally binding agreement between two or more parties, each of whom is committed to fulfilling their obligations [30]. Typically, contracts are enforced by a centralized legal entity. However, smart contracts replace trusted third parties or mediators between contracting parties, instead using code execution that is automatically verified by network nodes in a decentralized blockchain [26]. As a result, they enable transactions between untrusted parties without requiring direct contact, reliance on third parties, or intermediary fees [24].

A secret contract is a type of smart contract that facilitates transactions between two or more parties while keeping the agreement's contents hidden from others. The purpose of secret contracts is to safeguard the confidentiality and privacy of contractual arrangements. They are executed on a blockchain network, ensuring the security and transparency of the transaction. Secret contracts differ from conventional smart contracts in various aspects, including their distinct features. Some of these features include:

- **Confidentiality:** The primary goal of secret contracts is to uphold the confidentiality of the contractual agreement. Only the parties engaged in the agreement can view the contents of the contract.
- **Privacy:** By guaranteeing the confidentiality of the parties engaged in the agreement, secret contracts safeguard their privacy. The identities of the parties are kept hidden from others.
- **Security:** Secret contracts are implemented on a blockchain network, creating a secure and tamper-proof setting. The blockchain network guarantees that the contract's terms cannot be modified after execution.

Smart contracts are contracts whose terms and conditions are written in code, making them self-executing. They are executed on a blockchain network based on certain set of conditions, eliminating the need for intermediaries or third parties and enabling the automatic execution. Smart contracts are highly flexible and customizable which suit different use cases. Finance, supply chain management, real estate, and healthcare industries can adopt smart contract to automate processes, reduce costs, and increase efficiency. Figure 5 shows the steps in the process of smart contract execution.

Ethereum is the most popular smart contract platform. Solidity is the programming language for writing smart contract in Ethereum platform which is similar to Javascript language. EOS, TRON, and Hyperledger Fabric are some other smart contract platforms. Chaincode is the smart contract in Hyperledger platform. Industries can adopt different smart contract platform based on their needs and requirements.

Smart contract code may contain security vulnerabilities and bugs. Smart contract auditing is the process of reviewing and testing these code for identifying and fixing those vulnerabilities and bugs. Formal verification is the process of verifying the correctness of the smart contract code. It can help to reduce the risk of security vulnerabilities. Without having to manage the underlying blockchain infrastructure, there are some cloud-based services called smart contracts as a service (SCaaS) that allows organizations to deploy and manage smart contracts. Smart contract templates, testing smart contracts, deployment tools, security, and auditing are some services offered by these SCaaS service providers. AWS provides managed blockchain and smart contract services.

Participants (nodes) in a blockchain network has to solve complex problem for creating a transaction and block. Nodes get reward for solving problems and creating transaction and block. Reward are paid as gas fee. Gas fees are paid in the blockchain's native currency (e.g., ether for Ethereum). There are several challenges

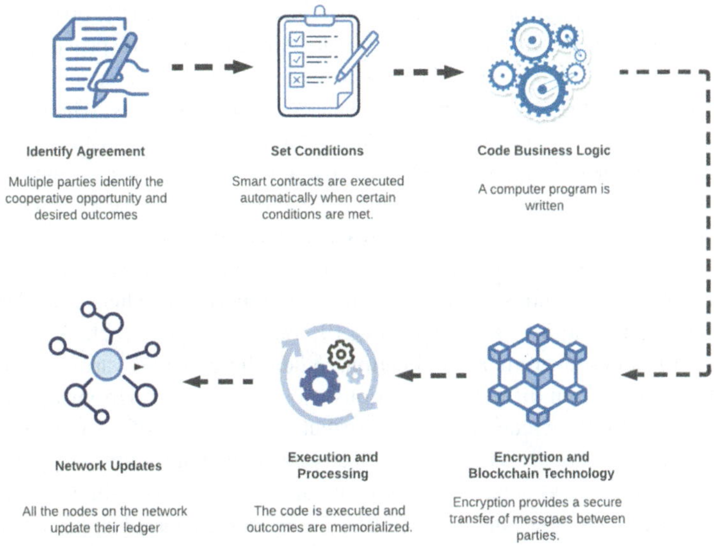

Fig. 5 How does smart contract work

in smart contract technology, such as code vulnerabilities, scalability issues, etc. Researchers and developers are working to address these challenges to make smart contracts more secure and scalable.

3 Related Work

The secure communication facilitated by blockchain-based zero-knowledge applications has gained significant attention in recent times. To emphasize the importance of privacy and security in government applications. Yang et al. [29] address the security and privacy challenges associated with e-government systems in smart cities in their research. It acknowledges the increasing vulnerability of sensitive online information and the constant evolution of sophisticated hacking techniques. The blockchain ensures the privacy, security, and integrity of sensitive information, while AI enables the development of intrusion detection and prevention systems. The implementation of zero-proof blockchain in real-world scenarios can enhance privacy and security measures. Johannes Sedlmeir, Micheal, and Egor explore the broader application possibilities of blockchain-based zero-proof protocols in various real-world contexts, such as e-commerce, green electricity, and enterprises in general [5, 10, 21]. Within the entire e-government network, blockchain technology provides a secure system where data is encrypted and distributed over the whole network that ensures information security and privacy. Elisa et al. [4] introduce a decentralized peer-to-peer (p2p) system for e-government, utilizing blockchain technology. This framework aims to enhance information security, protect privacy, and foster trust in public sectors. As blockchain ensures data privacy, security, and integrity, Makhdoom et al. [13] presented a data privacy scheme based on blockchain technology for secure data sharing in smart cities. Their approach involves dividing the blockchain network into multiple channels, with each channel comprising authorized organizations responsible for specific types of data, such as health, smart vehicles, smart energy, or financial details. To ensure security, users accessing the blockchain network are provided with dual protection in the form of API keys and OAuth 2.0, creating an isolation system from external sources.

Ensuring privacy and security is of utmost importance when considering e-government applications. Konstantinidis [9] has employed a blockchain-based approach for an e-government application. Where he explored various definitions of digital identity from different institutions and clarified the relationships between identities, identifiers, credentials, and attributes. The roles of identity providers (IdPs) and service providers (SPs) were discussed, along with centralized, federated, and decentralized approaches to identity management (IdM), while Auerbach et al. [27] discuss on how an anonymous digital identity-based approach can be implemented in e-government applications. On the other side, smart contracts have garnered significant attention in both academic and industrial circles as a hot research topic. These contracts possess decentralization, enforceability, and verifiability characteristics that enable the execution of contract terms between untrusted parties without the need for a trusted authority or central server. This potential has led to high expectations for smart contracts in revolutionizing various traditional industries such as finance, management, and the Internet of Things (IoT). Several survey research papers delve into the opportunities and prospects of implementing smart contracts in real-world applications, providing in-depth analysis and insights [14, 18, 25].

However, transaction data recorded and stored in the blockchain are secure against tampering, forgery, and deletion by malicious users. Due to the open and transparent nature of blockchain, privacy concerns still exist during transaction implementation. Therefore, an article by author Xiaoqiang Sun [23] focuses on studying the usage of zero-knowledge proofs (ZKP) in the blockchain environment. The paper survey begins by introducing the framework, models, and applications of ZKP. It then provides an overview of blockchain technology and proposes a ZKP framework specifically designed for the blockchain environment. Barros et al. [3] proposed a privacy-preserving vaccination pass system that incorporates self-sovereign identity, blockchain technology, and zero-knowledge proof. This system allows for the verification of users' vaccination proof while minimizing the disclosure of personal and health data. Panja et al. [17] introduced a secure end-to-end verifiable e-voting system. This system utilizes blockchain technology and a cloud server to store e-voting ballots. The authors integrated secure multi-party computation and non-interactive zero-knowledge (NIZK) proof techniques to publish the final tally without disclosing the individual tallies from direct-recording electronic machines.

4 Proposed E-Government Framework for Optimized Cybersecurity

Modern public governance stands at the cusp of a transformative era. Digital platforms, once a mere facilitator of governmental procedures, have now become the bedrock of public service delivery. As we grapple with the challenges of the Fourth Industrial Revolution, there is an urgent call for state-of-the-art e-government frameworks. This research elucidates an advanced model predicated on a consortium blockchain and elite cybersecurity techniques (Fig. 6).

4.1 Blockchain 3.0: Delving into the Consortium Blockchain Paradigm

The quintessential promise of blockchain—to bring about transparency, immutability, and decentralized control—is even more compelling in a governmental context. Our exploration of the third generation of this technology begins with a nuanced understanding of its architecture.

4.1.1 Network Hierarchies: The Equilibrium of Full Nodes and Light Clients

Full Nodes These entities are the bedrock of a blockchain network. By maintaining a comprehensive copy of the blockchain's ledger, they form the definitive check-

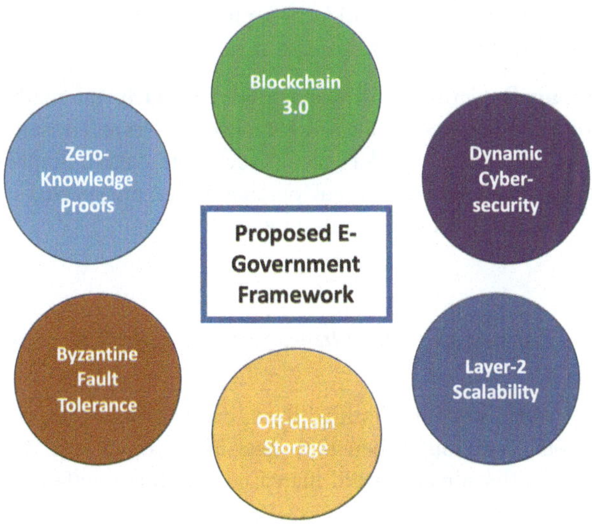

Fig. 6 Proposed e-government framework for optimized cybersecurity

points for data integrity. A proposed enhancement involves deploying dockerized environments to insulate and maintain node purity, thus preventing potential compromises. Coupled with the groundbreaking technique of sharding, where the database is partitioned to enable faster and concurrent transaction validation, full nodes present an optimized solution for large-scale government operations.

Light Clients For stakeholders who don't require the entirety of the ledger, light clients emerge as an agile alternative. Relying on Merkle-Patricia trees, these clients access and validate data sections pertinent to their interests, circumventing the need for exhaustive storage and computational capabilities.

The Cryptographic Pinnacle: ECDSA The cryptographic lynchpin of our proposition is the Elliptic Curve Digital Signature Algorithm (ECDSA). Renowned for its heightened security features, especially in contexts demanding limited computational power, ECDSA ensures transactional security, making unauthorized data alterations virtually impossible.

4.2 Dynamic Cybersecurity Framework: Marrying Traditional Strategies with Machine Intelligence

No technological advancement is devoid of challenges, and blockchain is no exception. Our framework, recognizing this, seamlessly amalgamates traditional cybersecurity practices with cutting-edge machine learning techniques.

External Defense Mechanism: Neural Network Armor Our defense strategy against external threats leverages the combined prowess of convolutional neural networks (CNNs) and long short-term memory (LSTM) networks. This amalgamation, trained on vast datasets of historical cyber threats, offers predictive capabilities, raising alerts well before potential breaches manifest.

Internal Anomaly Detection: A Dual-Faceted Approach Internally, the system's health is continuously gauged via a hybrid of autoencoders and dynamic complexity analysis (DCA). While autoencoders reconstruct input data to pinpoint anomalies, DCA scrutinizes the system's internal operations for unexpected computational complexities, shedding light on disguised internal threats.

4.3 Layer 2 Solutions: The Quest for Boundless Scalability

Blockchain networks, in their primitive form, can sometimes struggle with transaction throughput, especially when they face large volumes of data exchanges. Layer 2 solutions offer a path toward increased scalability without compromising on security.

State Channels These are essentially private lines of communication between parties, enabling off-chain transactions. Transactions within state channels do not need to be broadcasted to the entire network, making them significantly faster. Once the parties conclude their transactions, the final state is then broadcasted and recorded on the main blockchain. This reduces congestion and speeds up the transaction process.

Plasma Chains Inspired by the need for off-chain scalability, plasma chains, or child chains, operate in conjunction with the main blockchain (often referred to as the root chain). They handle local transactions independently, transmitting only essential data to the root chain. This hierarchical structure optimizes data flow and conserves network resources.

4.4 Off-Chain Storage: IPFS as the Torchbearer

The relentless growth in digital data necessitates advanced storage solutions. Traditional methods are ill-equipped to manage vast quantities of data without compromising on speed and accessibility.

InterPlanetary File System (IPFS) A protocol and network designed to create a peer-to-peer method of storing and sharing hypermedia. Unlike conventional data storage systems that retrieve data from a single location, IPFS retrieves data from multiple nodes, thus reducing the likelihood of a single point of failure. By harnessing content-addressable storage, IPFS ensures that data can be accessed even if certain nodes become unavailable, promoting a more resilient and decentralized storage ecosystem.

4.5 Privacy-Preserving User Interactions: The Brilliance of Zero-Knowledge Proofs

Data privacy remains one of the paramount concerns in the digital age. Ensuring user data confidentiality while maintaining the integrity of transactions is a challenge.

zk-SNARK (Zero-Knowledge Succinct Non-Interactive Argument of Knowledge) This cryptographic method allows one party to prove to another that a statement is true, without revealing any specific information about the statement itself. In the context of our e-government framework, zk-SNARKs can be deployed to verify transactions without disclosing their details, ensuring that user data remains confidential, while the integrity of transactions is maintained.

4.6 Ensuring Consensus: BFT Mechanism at the Core

Consensus mechanisms play a pivotal role in ensuring data consistency across a decentralized network. One of the most advanced consensus algorithms, especially in terms of fault tolerance, is the Byzantine fault tolerance (BFT).

Byzantine Fault Tolerance (BFT) BFT is designed to combat the Byzantine generals' problem, wherein nodes in a network must agree on a concerted strategy despite some nodes being unreliable or malicious. BFT ensures that as long as a pre-defined percentage of nodes are not malicious (typically 2/3rd), consensus can be achieved, ensuring data consistency and security. This level of resilience makes BFT a prime choice for our advanced e-government system, where data sanctity is of utmost importance.

4.7 Multi-Faceted Authentication: Embracing Biometric Modalities

As cyber-attacks grow sophisticated, multi-layered authentication mechanisms become crucial for robust security.

Iris Scans The unique patterns in the human iris provide an unparalleled authentication mechanism. Advanced computer vision algorithms can capture and compare these patterns, ensuring that access is granted only to verified individuals.

Minutiae-Based Fingerprint Recognition Fingerprint patterns contain unique features known as minutiae. Advanced biometric systems can map and compare these minutiae points, offering an efficient and secure authentication mechanism.

Facial Recognition Neural Networks (FRNN) Leveraging deep learning, FRNNs can analyze thousands of facial features to accurately recognize and authenticate individuals. Their self-learning capabilities ensure that they continually evolve, accommodating subtle changes in facial features over time.

5 User Registration and Authentication Process

Integration of blockchain technology in e-government services necessitates a robust user registration and authentication mechanism. Given blockchain's inherent cryptographic properties, the protocol focuses on securing user identities and facilitating a streamlined service access.

5.1 Algorithmic Approach to User Registration

Registration Algorithm delineates the user registration process. Specifically, the procedures from lines 1–9 in the algorithm underscore the essential blockchain properties of identity creation and cryptographic key management across all devices in the network.

Algorithm 1: Blockchain-based user registration

Require: Registration intent, Network Nodes P
Ensure: A registered entity z
  $(K_{\text{main}}, K_{\text{sub}})$ = generateCryptoKeys()
  $entityID \leftarrow$ generateUniqueID()
  $BlockAddr \leftarrow$ defineBlockchainAddr() + $(K_{\text{main}}, K_{\text{sub}})$
  $(entityID, K_{\text{sub}}) \leftarrow$ SecureStore$(entityID, K_{\text{sub}})$
  $BWallet \leftarrow$ formulateBlockchainWallet() + $(K_{\text{main}}, K_{\text{sub}})$
  for each $p \in P$ **do**
    disseminateWallet$(p, BWallet)$
  end for
  $z \leftarrow$ validateRegisteredEntity()

Upon registration, the user is not only empowered to transact within the e-government system but also benefits from enhanced access control to their records. This registration process optimizes the e-government system's efficiency, attributing clear responsibilities to every consortium blockchain node.

5.2 Transaction Initialization and Record Submission

Post-registration, when a user submits a record to the e-government network, it undergoes authentication to initialize the transaction. The block then undergoes versioning, broadcasted network-wide for validation, and subsequently dispatched to the user's blockchain address in the e-government consortium network.

Every record in the user's blockchain address encapsulates:

1. User ID
2. The record value (e.g., property registration)
3. Record identifier (e.g., property registration number)

In blockchain semantics, every data instance can be visualized as a virtual asset.

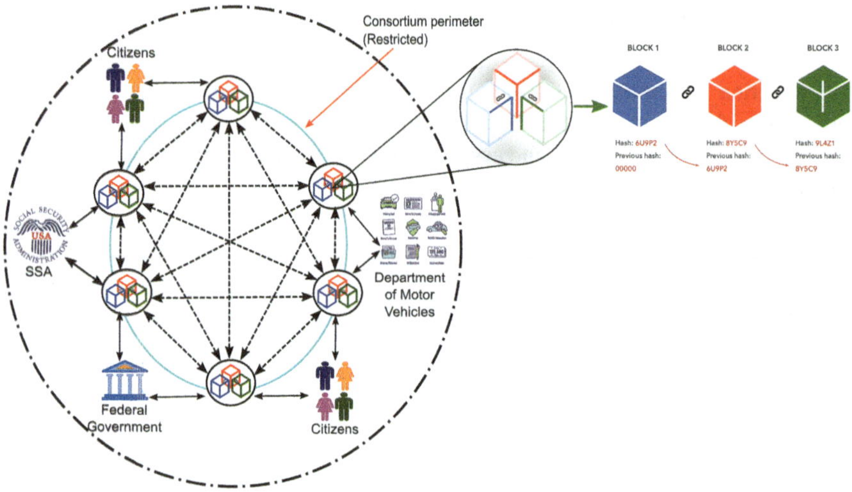

Fig. 7 Blockchain integrated e-government system

5.3 *Third-Party Access Mechanism*

For third-party entities to access a user's data, the user provides their blockchain address for verification. The external entity then utilizes the blockchain web API to fetch data stored in the user's blockchain address. It's imperative for users to securely manage their private keys. Loss of this key necessitates the creation of a fresh blockchain address, followed by a request to the e-government node to transfer old records to this new address (Fig. 7).

6 Identity Management and Authentication

Each user's identity undergoes stringent validation and authentication during network access via registered devices. Considering human errors as significant cybersecurity vulnerabilities, this two-tier authentication significantly curtails such potential lapses. Consequently, this ensures that government data is seamlessly and securely accessible to authorized individuals, unrestricted by their geographical location, provided Internet access.

6.1 *Blockchain-Based Identity Management*

Identity management plays a critical role in the digital era, where individuals interact with a multitude of online government platforms and services. However, centralized identity management systems often collect and store vast amounts of personal data, making them attractive targets for hackers. In recent years, blockchain technology has emerged as a potential solution to address these issues by providing a decentralized, transparent, and secure framework for identity management. Blockchain-based identity management shifts control to the individual, reducing the risk of data breaches and unauthorized access. The immutability of blockchain records ensures that tampering or manipulation of identity information is virtually impossible. Each transaction or update is recorded in a transparent and immutable manner, ensuring data integrity. It also provides individuals with greater control over their personal information. Users can selectively share their data and grant permission to entities, enhancing privacy and reducing the risk of identity theft. Blockchain-based identity management leverages cryptographic techniques, such as digital signatures, to ensure the authenticity and integrity of identity-related transactions. Digital signatures provide non-repudiation and proof of identity ownership. While blockchain ensures data integrity, it also introduces challenges in terms of privacy and consent management. Techniques such as zero-knowledge proofs, privacy-preserving smart contracts, and off-chain data storage-based standards and protocols can enable seamless integration and data exchange between different blockchain platforms. Moreover, blockchain scalability remains a challenge for identity management systems, as transaction volumes and verification requirements increase. Layer 2 solutions, interoperability protocols, and consensus algorithm optimizations are being explored to enhance scalability.

6.1.1 Blockchain-Based Identity Management Frameworks

Self-Sovereign Identity (SSI) SSI enables individuals to have full control over their digital identities, eliminating the need for intermediaries. Blockchain facilitates secure storage and management of personal identity attributes, allowing users to selectively disclose information for authentication without compromising privacy.

Decentralized Identifiers (DIDs) DIDs are unique identifiers anchored on a blockchain, providing a secure and globally resolvable identity mechanism. They enable verifiable and tamper-proof authentication by allowing users to prove ownership and control of their identifiers without relying on centralized identity providers.

Verifiable Credentials (VCs) VCs are digital credentials issued by trusted entities and anchored on the blockchain. They enable users to prove their identity or specific attributes without revealing unnecessary personal information. VCs are tamper-proof and can be verified by anyone with access to the blockchain.

Identity Wallets Identity wallets are user-centric applications that enable individuals to manage their digital identities and associated credentials securely. They provide a user-friendly interface for identity-related interactions, such as issuing, storing, and sharing verifiable credentials.

6.2 Blockchain-Based E-Government Authentication

Authentication is a critical step in verifying the identity of individuals, devices, or entities attempting to access government-related digital systems or resources. It plays a pivotal role in protecting sensitive information, securing transactions, and ensuring the integrity of digital interactions. However, traditional authentication methods have inherent limitations, such as weak passwords, susceptibility to phishing attacks, and centralized repositories that are attractive targets for hackers. Blockchain technology presents a novel approach to authentication by leveraging its decentralized, immutable, and transparent nature. Blockchain's immutability and cryptographic protocols make it highly resistant to hacking, data breaches, and identity theft. The decentralized and consensus-driven nature of blockchain ensures trust among participants, eliminating the need for a central authority. Users can maintain ownership and control over their identity attributes, selectively sharing information for authentication purposes while protecting their privacy. Blockchain's transparent nature enables audit trails and provides a verifiable record of authentication activities, enhancing accountability.

Public-Key Infrastructure (PKI) on Blockchain By storing public key certificates and associated credentials on a blockchain, PKI systems can leverage the immutability and transparency of blockchain for secure authentication. Blockchain-based PKI enhances the integrity and availability of digital certificates, reducing reliance on centralized certificate authorities.

In the next section, we delve into the potential applications and e-government services that can use our proposed framework.

7 Blockchain Integrated E-Government Services

We explore the applications of blockchain technology in e-government, including its benefits, challenges, and limitations. Specifically, we examine the use of blockchain technology in improving the efficiency, transparency, and security of e-government services. We also discuss the challenges associated with implementing blockchain technology in e-government, such as legal and regulatory barriers, interoperability issues, and the need for a skilled workforce.

- **Authentication:** Authentication is a critical aspect of secure digital systems, ensuring that users are who they claim to be and granting them appropriate access. Traditional authentication methods, such as passwords and two-factor authentication, are prone to vulnerabilities, including data breaches and identity theft. Due to its inherent security features, the potential of blockchain technology can act as a robust solution for authentication.
- **Identity Management:** Blockchain can be used to establish secure and decentralized digital identities for citizens. This enables individuals to have con-

trol over their personal information while facilitating seamless authentication and verification processes. Blockchain-based identity management systems can enhance security, reduce identity fraud, and streamline government services that require identity verification, such as issuing passports, driver's licenses, and social benefits.

- **Document and Record Management:** Blockchain technology can provide an immutable and transparent ledger for storing and managing government documents and records. By utilizing blockchain, government agencies can ensure the integrity, authenticity, and traceability of important documents, such as land titles, birth certificates, and business licenses. This eliminates the need for manual record-keeping, reduces the risk of data tampering, and simplifies the retrieval and verification of documents.
- **Supply Chain Management:** Blockchain can enhance transparency and efficiency in government procurement and supply chain processes. By recording the entire lifecycle of goods and services on the blockchain, governments can track and verify the origin, quality, and authenticity of products, ensuring compliance with regulations and combating counterfeiting. Smart contracts on the blockchain can automate payment processes, reduce administrative burdens, and facilitate secure and timely transactions between government entities and suppliers.
- **Voting Systems:** Blockchain-based e-voting systems can enhance the integrity, transparency, and accessibility of elections. By recording votes on a decentralized blockchain ledger, governments can ensure that votes are secure, tamper-resistant, and verifiable. Blockchain-based voting systems can prevent voter fraud, enable real-time vote counting, and enhance public trust in the electoral process. Additionally, blockchain can enable secure and anonymous voting, thereby safeguarding individual privacy.
- **Public Financial Management:** Blockchain technology can improve the efficiency, transparency, and accountability of government financial transactions. Through blockchain-based systems, governments can track and record financial transactions, budget allocations, and expenditure in a transparent manner. This enables real-time auditing, reduces the risk of corruption, and enhances public trust in the financial management of public funds.
- **Regulatory Compliance:** Blockchain can facilitate regulatory compliance by automating the monitoring and enforcement of regulations. By utilizing smart contracts, government agencies can ensure that businesses adhere to specific rules and regulations. For instance, blockchain can be used to monitor tax compliance, verify licenses and permits, and enforce regulations related to environmental protection or consumer safety.

8 Conclusion and Future Directions

This chapter has explored the potential of integrating blockchain technology into e-government systems to address some of the key challenges that these systems face, such as data security, transparency, and service delivery. Through an analysis of the core features of blockchain and real-world examples of its application in e-government, we have demonstrated the capacity of this technology to bring about significant improvements in the way public services are delivered and managed.

Blockchain's ability to provide a secure, transparent, and immutable record of transactions makes it an ideal tool for enhancing data security in e-government systems. The decentralized nature of blockchain also enhances transparency and accountability by allowing all transactions to be visible and auditable by all participants. In addition, by eliminating the need for intermediaries, blockchain can lead to more efficient service delivery.

However, the integration of blockchain into e-government systems is not without its challenges. These range from technical issues such as scalability and interoperability to non-technical issues such as the need for legislative and regulatory changes and the acceptance of the technology by the public. To overcome these challenges, a comprehensive and strategic approach that includes stakeholder involvement, technical innovation, and regulatory reform will be required.

The conceptual model we presented provides a blueprint for the integration of blockchain into e-government systems, outlining the key components and interactions that could underpin such a system. We believe that this model can serve as a starting point for government entities looking to harness the benefits of blockchain technology.

Looking to the future, more research is needed to refine the model and to explore how blockchain can be effectively and sustainably integrated into different e-government contexts. Issues such as the environmental impact of blockchain, the trade-offs between transparency and privacy, and the socioeconomic implications of blockchain implementation in e-government also warrant further exploration.

In conclusion, while blockchain presents several challenges, it holds immense potential to revolutionize e-government systems, ushering in a new era of secure, transparent, and efficient public service delivery. As we continue to grapple with the challenges of digital transformation in public administration, the integration of blockchain into e-government systems should be seriously considered as a viable strategy for future-proofing our public services.

References

1. Androulaki, E., Barger, A., Bortnikov, V., Cachin, C., Christidis, K., De Caro, A., Enyeart, D., Ferris, C., Laventman, G., Manevich, Y., et al.: Hyperledger fabric: a distributed operating system for permissioned blockchains. In: Proceedings of the thirteenth EuroSys conference, pp. 1–15 (2018)
2. B. Rawat, D., Chaudhary, V., Doku, R.: Blockchain technology: emerging applications and use cases for secure and trustworthy smart systems. Journal of Cybersecurity and Privacy **1**(1), 4–18 (2020)
3. Barros, M., Schardong, F., Cust'odio, R.: Leveraging self-sovereign identity, blockchain, and zero-knowledge proof to build a privacy-preserving vaccination pass (2022). https://doi.org/10.2139/SSRN.4036226

4. Elisa, N., Yang, L., Chao, F., Cao, Y.: A framework of blockchain-based secure and privacy-preserving e-government system. arXiv: Cryptography and Security (2020). https://doi.org/10.1007/S11276-018-1883-0
5. Ermolaev, E., Abellán Álvarez, I., Sedlmeir, J., Fridgen, G.: z-commerce: designing a data-minimizing one-click checkout solution. In: Proceedings of the Design Science Research for a New Society: Society 5.0: 18th International Conference on Design Science Research in Information Systems and Technology, DESRIST 2023, Pretoria, South Africa, May 31–June 2, 2023, pp. 3–17. Springer, Berlin (2023)
6. Foundation, T.L.: Hyperledger foundation (2022). https://www.hyperledger.org/
7. Iredale, G.: Introduction to blockchain features. 101 Blockchains (2021). https://101blockchains.com/introduction-to-blockchain-features/
8. Jakobsson, M., Juels, A.: Proofs of work and bread pudding protocols. In: Secure Information Networks: Communications and Multimedia Security IFIP TC6/TC11 Joint Working Conference on Communications and Multimedia Security (CMS'99) September 20–21, 1999, Leuven, Belgium, pp. 258–272. Springer, Berlin (1999)
9. Konstantinidis, G.: Identity and access management for e-government services in the european union–state of the art review (2021)
10. Kuperberg, M.: Blockchain-based identity management: a survey from the enterprise and ecosystem perspective. IEEE Trans Eng Manag **67**(4), 1008–1027 (2019)
11. Larimer, D.: Delegated proof-of-stake (DPOS). Bitshare Whitepaper **81**, 85 (2014)
12. Luu, L., Narayanan, V., Zheng, C., Baweja, K., Gilbert, S., Saxena, P.: A secure sharding protocol for open blockchains. In: Proceedings of the 2016 ACM SIGSAC Conference on Computer and Communications Security, pp. 17–30 (2016)
13. Makhdoom, I., Zhou, I., Abolhasan, M., Lipman, J., Ni, W.: Privysharing: a blockchain-based framework for privacy-preserving and secure data sharing in smart cities (2020). https://doi.org/10.1016/J.COSE.2019.101653
14. Morais, E., Koens, T., Van Wijk, C., Koren, A.: A survey on zero knowledge range proofs and applications. SN Appl. Sci. **1**, 1–17 (2019)
15. Nakamoto, S.: Bitcoin: A peer-to-peer electronic cash system. In: Decentralized Business Review, p. 21260 (2008)
16. Olson, K., Bowman, M., Mitchell, J., Amundson, S., Middleton, D., Montgomery, C.: Sawtooth: An Introduction. The Linux Foundation, San Francisco (2018)
17. Panja, S., Roy, B.: A secure end-to-end verifiable e-voting system using zero knowledge based blockchain (2018)
18. Partala, J., Nguyen, T.H., Pirttikangas, S.: Non-interactive zero-knowledge for blockchain: A survey. IEEE Access **8**, 227945–227961 (2020)
19. Ream, J., Chu, Y., Schatsky, D.: Upgrading blockchains. In: Deloitte Insights (2018)
20. Saleh, F.: Blockchain without waste: Proof-of-stake. Rev. Financ. Stud. **34**(3), 1156–1190 (2021)
21. Sedlmeir, J., Völter, F., Strüker, J.: The next stage of green electricity labeling: using zero-knowledge proofs for blockchain-based certificates of origin and use. ACM SIGENERGY Energy Informatics Review **1**(1), 20–31 (2021)
22. Service, A.W.: Amazon managed blockchain (2023). https://aws.amazon.com/managed-blockchain/
23. Sun, X., Yu, F.R., Zhang, P., Sun, Z., Xie, W., Peng, X.: A survey on zero-knowledge proof in blockchain. IEEE Netw. **35**(4), 198–205 (2021)
24. Swan, M.: Blockchain: Blueprint for a New Economy. O'Reilly Media, Inc., New York (2015)
25. Wang, S., Ouyang, L., Yuan, Y., Ni, X., Han, X., Wang, F.Y.: Blockchain-enabled smart contracts: architecture, applications, and future trends. IEEE Trans. Syst. Man Cybern.: Syst. **49**(11), 2266–2277 (2019)
26. Wang, S., Yuan, Y., Wang, X., Li, J., Qin, R., Wang, F.Y.: An overview of smart contract: architecture, applications, and future trends. In: 2018 IEEE Intelligent Vehicles Symposium (IV), pp. 108–113. IEEE, New York (2018)

27. Wirtschaftswissenschaftlichen, D., Auerbach, N.: Anonymous digital identity in e-government (2004)
28. Wood, G., et al.: Ethereum: a secure decentralised generalised transaction ledger. Ethereum Project Yellow Paper **151**(2014), 1–32 (2014)
29. Yang, L., Elisa, N., Eliot, N.: Privacy and security aspects of e-government in smart cities. In: Smart Cities Cybersecurity and Privacy, pp. 89–102. Elsevier, Amsterdam (2019)
30. Zheng, Z., Xie, S., Dai, H.N., Chen, W., Chen, X., Weng, J., Imran, M.: An overview on smart contracts: challenges, advances and platforms. Futur. Gener. Comput. Syst. **105**, 475–491 (2020)

Part III
Securing IoT with Blockchain

A Blockchain-Enabled Serverless Security Mechanism for IoT-Based Drones

Mohsen Ghorbian and Mostafa Ghobaei-Arani

Abstract The unique features of serverless computing have contributed to its popularity and widespread use. Despite these advantages and features, serverless computing technology faces challenges. Among these significant challenges are concerns over security and privacy protection. Providing a security system that could offer extremely high and safe security has always been something that has been considered. Blockchain technology can help make a safe authentication system for serverless computing because of its unique design and complex and powerful encryption methods. This chapter presents a method of authenticating serverless service provider systems that use permission-based blockchain technology on Hyperledger Fabric. This method offers highly secure levels of security. In the drone scenario, there are two parts of the security mechanism for serverless computing security with Hyperledger Fabric (SCSHF): registration in serverless computing security with Hyperledger Fabric (RSCSHF) and authentication in serverless computing security with Hyperledger Fabric (ASCSHF). This approach suggests using the blockchain/s features to create a highly secure authentication system, making it less likely that services and client data would be compromised. Using this approach, security concerns about protecting privacy and preventing unauthorized access to service systems, particularly in sensitive technologies like drones, will be addressed, ensuring the system's security.

Keywords Serverless computing · Security mechanism · Blockchain · Hyperledger Fabric · IoT-based drones · Authentication mechanism

M. Ghorbian · M. Ghobaei-Arani (✉)
Department of Computer Engineering, Qom Branch, Islamic Azad University, Qom, Iran
e-mail: mo.ghobaei@iau.ac.ir

N. El Madhoun et al. (eds.), *Building Cybersecurity Applications with Blockchain and Smart Contracts*, Signals and Communication Technology,
https://doi.org/10.1007/978-3-031-50733-5_3

1 Introduction

The cloud computing infrastructure and environment can be categorized into four main types based on the services provided and the locations of the service providers. These types include private, public, community, and hybrid clouds [1]. Generally, the cloud computing services model provided can be divided into three categories based on their type of service: IaaS (infrastructure as a service), PaaS (platform as a service), and SaaS (software as a service). Due to this, these core services are the basis for most other services [2]. In recent years, a new service model in cloud computing has emerged, known as FaaS (function-as-a-service). Since this service delivery model is relatively new compared to other existing service delivery models, the place it belongs in comparison to other existing models is among the PaaS and SaaS delivery models. This service model allows developers to develop their programs most quickly and efficiently as possible. Understanding the concept of serverless computing is necessary to use the service model. The term "serverless computing" is used in some reliable and official sources to refer to the mechanism for providing services based on FaaS. Despite this, it is essential to note that the serverless computing model has made it feasible to provide functions-based services due to the architecture it uses in its implementation process. Using serverless computing, developers do not have to constantly manage their servers and infrastructure, which allows them to build apps quickly and easily [3]. In other words, a developer will not be responsible for setting up, planning, or managing servers, resources, and infrastructure for projects. Therefore, to accomplish their processing tasks, clients must call and execute the functions in this architecture to complete their assignments on time. The serverless architecture is highly scalable and offers an accurate and new payment model. The new model allows clients only to pay the cost of the resources they utilize. In other words, clients do not need to pay for resources they do not use. In another hand, it is an event-based mechanism, which means that a particular event can trigger its functions. Hence, it is called an event-based mechanism [4]. The issue of security is always of utmost importance when it comes to structures such as serverless computing since it is a technology that interacts with clients continuously and consistently. There is also the matter of privacy which will be important to clients since they would like the reassurance that their information is protected from unauthorized access and that their privacy is always protected. On the other hand, there have been systems that have always tried to implement solutions that can eliminate their clients' concerns about their privacy and have always done their best to protect clients' privacy from unauthorized access. Hence, a provider system tries to prevent untrustworthy people from accessing it and limits access to them by putting all their effort into being used. It is, therefore, essential to provide a solution and adopt an approach through which clients can have confidence that their information and, in other words, the privacy they have defined for themselves will be protected by a system when they use the services of a service provider [5]. In this way, people seeking abuse will not be able to abuse other clients' personal information. It is also imperative that service providers ensure that unreliable and unapproved individuals will not be able to access and utilize the system's services. Ultimately, the blockchain will be able to address the clients' and providers' concerns and needs [6]. Due to their innovative security approach, blockchains have always been known as pioneers in providing security and preventing unauthorized access to information. These mechanisms, with the benefit of powerful application techniques and methodologies, always provide the possibility that illegally accessing information will be lost. They owe this advantage to the block creation approach, consensus mechanism, and utilization of powerful cryptographic algorithms. Blockchain technology can be classified into public and private, each with its application and purpose [7]. Serverless computing services are presented to clients in a manner that requires them to confirm their identity before requesting services from the requester. Once their identity has been verified, they will be given the possibility to submit their request. Serverless computing authentication mechanisms are always susceptible to vulnerabilities. Suppose a client intends to abuse the system. In that case, they may have access to the information of clients that have already registered, and by entering their information into this system, they can prove themselves as authenticated clients and perform their actions [8]. By utilizing the blockchain approach for registering and authenticating clients before accessing the system, the service providers and clients can ensure that their information will not be misused at any time. Furthermore, the system will ensure that only clients the system has approved will have access to it and its services. Therefore, the proposed mechanism in this chapter provides a security-based issue solution using serverless computing technology based on blockchain technology to provide security in communication between clients and provider systems using serverless computing technology. A drone scenario is used to illustrate the proposed mechanism. The approach proposed in this scenario demonstrates how a serverless computing service provider utilizing blockchain technology attempts to protect itself and its clients from unauthorized access by ensuring the security of both parties. It is also possible that the system can resist most typical attacks executed by abusing the system and does not affect it.

This chapter is structured as follows: As a starting point, the first section examines the essential prerequisites, such as explaining serverless computing with its advantages and disadvantages, blockchain technology, and drone technology in complete detail. The second section of this chapter describes drone security with blockchain in serverless computing, serverless computing architecture, and blockchain technology based on Hyperledger Fabric. The third section examines several important security challenges in serverless computing. In the fourth section of this chapter, a hybrid structure for serverless computing security with Hyperledger Fabric (SCSHF), such as registration in serverless computing security with Hyperledger Fabric (RSCSHF) and authentication in serverless computing security with Hyperledger Fabric (ASCSHF), is proposed in the drones' fields. Finally, it is concluded in the fifth section.

2 Background

This section will provide a concise overview of serverless computing, blockchain technology, drone structures, and the most important security parameters to consider.

2.1 *Serverless Computing*

The serverless computing approach enables the development of new service models by implementing the microservice architecture. It involves the execution of code blocks known as functions that provide the services requested by the client [9]. Hence, these code blocks are called executable programs and are responsible for providing the requested services. An application with a monolithic architecture that offers various services is converted into several small programs known as functions using a serverless computing architecture based on a microservice architecture [10]. In this application, all of the services that are provided by it will be transformed into several small programs called functions, thereby making the application more efficient. In this manner, the program with monolithic architecture with all services converts to functions that will run independently for each service [11]. Therefore, for the presentation of a service, serverless computing services utilize FaaS (functions as a service), a model based on function-as-a-service. A key component of serverless computing is FaaS (functions as a service), which cloud providers provide in a scalable, fully managed manner and stateless manner. Therefore, this technology makes it possible for devices that lack enough resources to run the required programs and services by utilizing this feature. The code that is mentioned under the program name can be classified into a variety of categories. For example, computational functions are functions whose primary purpose is to perform processing tasks. The characteristic of serverless computing functions is that they are defined as stateless. The definition of stateless functions implies that services presented in the code do not need data for processing. In other words, if a device wants to run a program based on a monolithic architecture, the device will also need to receive and execute a large amount of data related to that program along with receiving and executing the program codes [12]. In serverless computing, this can be solved by converting this program into parts that do not require the data defined in the program to run. In this regard, it is crucial to note that when a program is converted into stateless codes, the execution process of these programs becomes much faster. Therefore, the clients will be able to consume fewer resources and ultimately pay a lower amount [13]. Serverless computing offers many advantages in a wide range of fields. The following are a few of those advantages:

- *Rapid development*: Serverless computing aims to keep clients as far away from the complex processes associated with infrastructure configuration as possible. Hence, this allows clients, specifically developers, to concentrate on developing, testing, and deploying their applications. Therefore, this feature allows the development of applications to be much faster and easier. Due to the use of this feature, the process of implementing programs will be more rapid [14].

- *Providing accurate cost calculation models*: This feature is undoubtedly one of the most significant advantages of serverless computing. According to this calculation, the model called pay-per-use, clients will be paying costs according to the amount of resources consumed [15]. In other models, such as pay-as-you-go, the client would be charged based on the time the resource was rented. For example, in this model, the client rents resources for some time but only use a portion of the time he has access to them. In this regard, the client is liable for all costs incurred, even when resources were not utilized. Consequently, this feature of serverless computing can also be helpful for clients because clients pay for the number of resources they use. In addition, this can also be helpful for service providers because they can use essentially unused resources to provide services to additional clients [14].
- *Framework*: Various frameworks are available for the implementation of the serverless concept. Consequently, developers have many choices when it comes to serverless frameworks. Several frameworks are available in the current world of serverless architecture that can help developers build, package, compile, and deploy their code. Building, packaging, compiling, and deploying code in the cloud takes minutes with these frameworks. The coding process is simplified through these frameworks, and the scaling process simplifies the whole process. These frameworks include OpenWhisk, OpenFaaS, Apex, Google Cloud Functions, and AWS Serverless Application platform [16].
- *Ability to run programs on tiny devices*: Serverless computing has enabled providers to provide services as functions. Therefore, implementing these functions usually requires only a limited amount of resources. Consequently, by using this feature, devices typically located at the edge of a network and are limited in resources are also able to receive their requested services quickly, i.e., be able to run the required programs on the device [16].
- *Programming language support*: Serverless computing services and the creation of functions require using platforms. Several platforms have been developed to provide serverless services, some of which are open source and others are closed source, which large companies generally provide. Thus, all these platforms should allow developers to create the functions they need based on the programming language in which they are proficient, to meet the needs of clients who are generally developers in this field. Hence, the platforms available must support powerful and famous languages such as Python, Node.js, Go, Java, and JavaScript. Still, there may be differences in the languages they support between various platforms [16].

Generally, new technologies have been developed to solve current problems and provide tools to assist in solving future problems. Due to their built-in capabilities, these technologies provide the necessary capabilities to solve problems efficiently and effectively [17]. Hence, serverless computing as a new technology is still in its infancy and will inevitably have disadvantages. Therefore, there are some disadvantages associated with this technology, including:

- The services and responsibilities of each function are assigned to accomplish a specific task, so serverless providers should ensure that their clients can access what they need. Hence, it is crucial to configure permissions and rights correctly to prevent functions from becoming overprivileged, posing a security risk.
- In serverless architecture, it is often difficult to debug an application or function from line to line; some developers prefer to turn on debugging mode and enable frequent error reports to simplify the debugging process. Hence, during the transition to production, developers may neglect to clean the reports, which results in verbose error messages. Consequently, this may reveal information about the serverless functions and their logic that this mistake can take high risks.
- There are multiple dependencies in serverless applications, including those that integrate with databases, cloud services, and various other components. Consequently, if these dependencies contain vulnerabilities, they can also be exploited by attackers. The cloud service provider's responsibility is to ensure the security of all cloud components, including the data center, the network, the servers, the operating system, and their configurations. It is also possible for the developer to improve security.
- When a function instance is started for the first time, it is impossible to retrieve stateful information from previous instances. As a result, when a new instance of the function is launched for the first time, it does not retrieve the stateful information from previous instances [18].

A system will always face challenges and issues related to privacy and security. Safety and security are fundamental elements of any successful and effective system. This issue also applies to systems that use serverless computing to provide services since these systems always experience the same problem. Despite some successes in finding solutions for this particular problem, it is essential to note that it remains an open issue in this field, which should be explored further.

2.2 *Blockchain*

Blockchain is a distributed database consisting of sequential blocks of information stored on a distributed network and immutably online. Information contained in each block is coded (Hash) from the previous block, and new information is added. This information may be relevant to digital transactions, smart contracts, digital assets, or any other type of information. The blockchain can be divided into two types: permissionless and permission-based [19].

Permissionless Systems

Permissionless systems allow anyone to access the network and conduct transactions. They do not require permission to access the network, and no one controls it. Therefore, anyone with a computer and an Internet connection can participate in the network. A permissionless system provides high security and transparency, as no one can alter the data recorded in the network. It is important to note that unlicensed systems may be vulnerable to security attacks due to the breach of some security barriers. Bitcoin and Ethereum are examples of permissionless systems. It is common for permissionless blockchain-based systems to be used in industries such as digital currency, e-commerce, and health, which require high levels of security and transparency. As a result of the large number of users and small- and medium-sized enterprises using these systems, there are no restrictions on access to networks and data. These systems are ideally suited for areas that require a high level of security and transparency because of their high level of security and transparency. Permissionless systems have both advantages and disadvantages. The advantages include:

- *Lack of license dependence*: Permissionless systems require no license to access the network and data, and anyone can participate.
- *High transparency*: Permissionless system ensures high transparency by making all transactions public.
- *Lower cost*: Permissionless system is less expensive since it requires less technical expertise and fewer servers.

There are several disadvantages associated with permissionless systems, including:

- *Lower security*: Permissionless systems may be deprecated and vulnerable to security threats due to the lack of authorization and greater control.
- *Higher risk*: Permissionless systems may have higher risks due to a lack of control and management. For example, in digital currencies, there is a possibility of losing currencies.
- *Illegality*: Some permissionless systems are illegal due to noncompliance with regulations and laws and may face legal problems [20].

Permission-Based Systems

Permission-based systems allow authorized users to access a network and perform transactions. As a result, only individuals granted access to the network and its data by the network administrators are permitted to operate on the network. Network administrators can also control and make decisions regarding the network. In licensed systems, security and efficiency are higher than in unlicensed systems. Permissioned systems, such as Hyperledger Fabric and Corda, are more expensive than open-source systems due to the complexity of setting them up and maintaining them. Permission-based systems are commonly used in banking and insurance areas requiring strict control and management. Large companies, banks, industries, and governments generally use these systems. Only authorized persons can access the network and data in these systems. Due to their greater security and control, these systems are generally suitable for areas requiring high precision and accuracy.

It is important to note that permission-based systems have both advantages and disadvantages. The advantages include:

- *Higher security*: A permission-based system provides greater security than an unlicensed system because the access permissions are stricter.
- *Control and management*: A permission-based system allows licensees to control and manage the network, which makes it easier for them to solve problems.
- *Proprietary rights protection*: Permission-based systems provide greater control and management over sensitive data and information.

There are several disadvantages associated with permission-based systems, including:

- *Greater complexity*: Permission-based system requires more technical expertise to install and configure.
- *Higher cost*: Permission-based system is more expensive as it requires more technical expertise and powerful servers.
- *More control*: Permission-based system provides greater control, which may limit the maps already created [21].

2.3 Drones

Uncrewed aerial vehicles (UAVs), also known as drones, can be used for various purposes. From simple to complex tasks, UAVs can be used. They can assist rescue victims of disasters such as flooding, mountain avalanches, earthquakes, etc. They can also deliver packages by mail. In other words, uncrewed aerial vehicles are robotic devices initially developed for military and aerospace applications. Due to their relative safety and efficiency, drones have become increasingly popular as a product of new technology. There are a variety of levels of autonomy available. Drones can achieve various levels of autonomy, including remote piloted (where a human controls the drone) and advanced autonomy (where the sensors and LiDAR sensors are utilized to calculate the drone's movement). Different types of drones can travel at different heights and distances. For example, a close-range uncrewed aerial vehicle (UAV) can travel 30 miles from its base. Surveillance and intelligence gathering are most commonly performed by UAVs that travel a short distance or can travel up to 100 miles. In addition to intelligence gathering and scientific studies, mid-range UAVs can conduct meteorological research, scientific research, and scientific studies. It has a range of 500 miles and can fly up to 3200 feet in the air, making it an endurance drone. As drones can be remotely controlled and flown to different distances and heights, drones are excellent candidates for some of the most challenging jobs in the world [22]. When selecting drones, pilots need to consider various factors, including the ability to deliver commands and the device's overall weight. Ground control stations, payloads, and data links are three essential components of drones. In the following sections, each case will be discussed in detail.

- *Payloads*: The importance of drones cannot be overstated since they are available in a wide range of sizes and a wide variety of payloads, which can also be variable in size and carried by a wide range of drones. It is important to note that drone deliveries can be used for various purposes, ranging from lifesaving medications and packages to more. Still, they must be constructed to enable them to perform efficiently and efficiently at a reasonable cost.
- *Ground Control Station (GCS)*: The ground control station is responsible for controlling the flight and operation of UAVs, while the central control unit tracks the movements of the UAVs. These devices come in various sizes, depending on their size and shape. These devices can be tiny to fit on a desk, in a handheld controller, or very large, depending on their size and weight. In addition to controlling flight, payload sensors, displaying status readings, and planning missions remotely through satellites, a GCS system enables its clients to control flight, payload sensors, and provide status readings.
- *Data Links*: During the flight, the data link serves as a communication center between the drone and the ground operator, connecting the drone to the ground operator. The data link also provides the operator with airspeed altitude and remaining flight time, speed, and distance from the target. When the operator uses 2.4 GHz for video and 5 GHz for control, the range of the UAV of video and control is approximately 4 miles [23].

Types of Drones Based on Wing Type

Different industries require different types of drones, each tailored to meet the requirements of their particular industry. For instance, some industries use lightweight drones to transport cameras for photography, while others require robust drones to carry heavy medical supplies. Therefore, many companies are manufacturing four types of drones to meet these needs—single-rotor helicopter drones, fixed-wing drones, multirotor drones, and fixed-wing hybrid VTOL drones.

- *Single-rotor helicopter drones*: A single-rotor helicopter drone looks similar to a helicopter in design and construction, with just a single rotor providing power and a tail controlling direction. In order to carry more payload and fly more efficiently than multirotor drones, this drone combines the advantages of tiny multirotor drones with single-rotor drones.
- *Fixed-wing hybrid VTOL drones*: A fixed-wing hybrid VTOL drone is the new drone technology. It is a plane with fixed wings that takes off and lands vertically. Besides offering long ranges and flight times, fixed-wing UAVs also provide rotary-wing devices with vertical takeoff capabilities, eliminating the drawbacks associated with fixed-wing devices, which require a great deal of space when taking off and landing [24].
- *Multirotor drones*: A multirotor drone, also known as a rotary-wing drone, is widely used in agriculture and for commercial and recreational purposes. In

addition to their small size, excellent control, and good visibility, multirotor drones are excellent aerial photography tools.

- *Fixed-wing drones*: The design and appearance of fixed-wing drones are similar to radio-controlled airplanes, so they are commonly used for mapping large areas and harnessing wind power. Their drone's aerodynamic design enables them to hover longer and travel faster and farther, thanks to their streamlined design. On the other hand, fixed-wing drones tend to be more expensive than multirotor drones since they have a fixed-wing and a multirotor [25].

3 Drone Security with Blockchain in Serverless Computing

In addition to providing exceptional security to its clients, a blockchain network has another unique feature called transparency. This unique characteristic has typically been recognized in blockchain networks. By using blockchain technology, network clients can secure their exchanges using powerful encryption features provided by the algorithms in blockchain technology. All transactions are encrypted before they are stored by powerful encryption algorithms, allowing their security policies to be implemented and their security to be guaranteed. It is possible to use blockchain technology for various purposes, but privacy concerns keep many businesses from adopting it. Therefore, blockchain technology should have features that can address privacy concerns and allow the use of this technology for specific applications by the purposes for which these applications are used. Hence, this technology should be used so that clients can plan and adjust it quickly and according to the defined goals and ultimately guarantee their privacy and security [26]. The advantages of serverless computing include the availability of a wide range of services suitable for developing applications in various industries. The architecture used in this technology allows developers to develop and run programs with this type of computing at a lower speed and cost because this architecture is a type of microservice architecture. By converting cloud applications into codes called functions, clients can call and execute only the services they need in the form of stateless functions instead of complete applications, which is one of the main advantages of this approach. In addition, it is possible to call and use one or more functions simultaneously, which is a great advantage. For clients to utilize the provided services in serverless computing, the authentication process must first be applied to them, ensuring the system ensures that the clients who will be using the provided services have authenticated themselves. In other words, the identities of the clients have been verified [27]. These computing services can prove extremely useful in drone missions. They can reduce their energy consumption by using serverless services in drone missions. For example, drones are equipped with sensors that detect movements on the earth's surface and radiation related to substances such as radioactivity. These drones can also be used to take pictures of the environment. By using serverless computing in drones, functions can be called and executed to process each of the cases mentioned, thereby significantly reducing the energy consumed by these drones. In this chapter, the proposed plan tries to turn the weakness of serverless computing systems' authentication process into its strength by using blockchain technology. Doing so makes it practically impossible to alter the authentication and impersonation processes to access serverless service providers' services. For example, drones must authenticate to the serverless service providers' systems before they can access serverless services. This issue may be problematic if a drone attempts to access the system through network penetration or impersonation. In order to overcome this problem and challenge, the upcoming approach uses the facilities provided by a blockchain. A serverless service provider system can use the considered mechanism to authenticate drones and store their information in a blockchain, allowing it to obtain information about drones whenever a drone requests to use the serverless service. In the next step, we describe how Hyperledger Fabric's private blockchain works with serverless computing.

3.1 Serverless Computing Approach

Function-as-a-service and backend-as-a-service have been significant concepts in the development of serverless systems. A serverless system requires several components, including a storage system, database, framework, technology stack, etc. In event-driven code execution systems, applications are executed in response to requests, similarly managed by cloud platforms to function-as-a-service. Based on events, cloud platforms deploy logic in software containers and execute it on demand. As opposed to the backend-as-a-service, function-as-a-service allows developers to build custom applications without relying on libraries with prewritten code. The API gateway enables cloud providers to manage serverless applications, which developers can access directly via APIs, simplifying data integration, running for shorter periods, and being stateless [28]. It is important to note that backend-as-a-service and function-as-a-service share many similarities because both require third-party service providers to provide services. As a result of the backend-as-a-service model, developers can concentrate on writing front-end code, while backend services, such as storage, are provided. Backend-as-a-service applications are not event-driven like serverless applications and may not run at the edge. Backend-as-a-service is offered, for example, by Amazon Lambda. Lambda provides guidelines for developers to follow when submitting serverless code to containers, automates the process of putting code into containers, and offers managed services to developers. The ability to construct a successful serverless application requires a solid understanding of the logical structures that these architectures come with, which must be understood to ensure their proper operation. For serverless architecture to function correctly, you must understand the logical frameworks in which these architectures come packaged [29]. Within this framework, several components contribute to developing logical systems and applications. All of these components are essential for the development of such systems and applications. Understanding that many components are involved in this framework is crucial for success. All

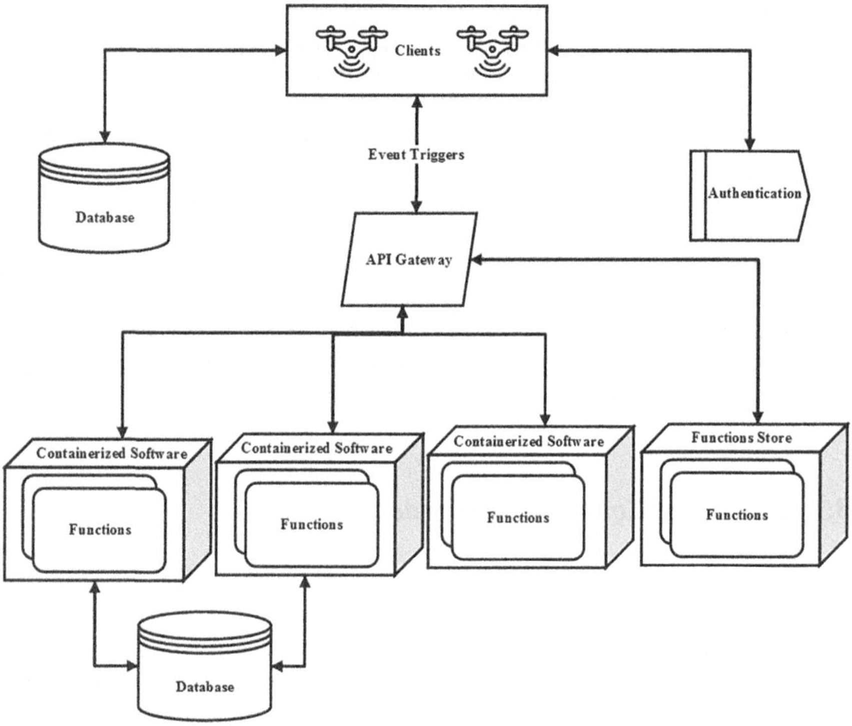

Fig. 1 Serverless mechanism

architectural components are shown in Fig. 1. Architectures based on serverless computing include the following components:

- *Clients*: The client is integral to serverless architectures, so they cannot be forced into any application. They must handle small bursts of requests in a stateless manner, as well as integrate with a variety of platforms flexibly. Moreover, the interface must handle both exceptionally high and shallow volumes of data simultaneously.
- *Authentication*: The process of authorizing a client to use a particular resource in a system security system is called an access client or control privilege in system security. Authorization always precedes authentication in secure environments. Before organization administrators can grant clients access to requested resources, they must verify that their credentials are confirmed. Serverless computing systems will follow the same process. Clients who wish to utilize the services provided by these systems must first confirm their identity and possess authorization to use them [30].
- *Containerized Software*: A serverless architecture enables providers to deliver containerized software with microservice architecture without complexity. Their platform is made available to developers through a containerized microservice

architecture. They also serve as an integrated code repository, allowing developers to develop code for multiple platforms, including desktops and smart devices, without having to deal with complicated resource management infrastructure [15].

- *Databases*: The functions use a database to access data and interact with other functions. The most important thing to consider is that these databases can provide serverless functions and store data for an indefinite period, which is extremely important to consider. Several serverless databases are available today, and many are publicly accessible. Aurora Serverless, Azure Cosmos DB, DynamoDB, and Cloud Firestore are just a few examples of the services currently available to the general public.
- *API Gateway*: In a serverless API gateway, client calls are routed to the appropriate microservice within the application to make seamless communication between the applications possible. Client calls are routed to the appropriate microservice within the application. This software can create, publish, monitor, maintain, and secure a REST API or web service. Several benefits and features are offered by API gateway to fulfill its core mission of facilitating API requests, including effective client service, communication with partners and suppliers, and integration with internal systems [31].
- *Event Triggers*: Serverless framework offers a service that defines a public HTTP endpoint for serverless functions. API gateways are also included in serverless platforms. Using the serverless framework's HTTP events syntax, HTTP endpoints can be used as event sources. In serverless architectures, functions serve as routes and endpoints mapped to the configuration resources. API gateways invoke functions when they receive a request when the routing configuration matches the request [32].
- *Functions Store*: As part of serverless computing, function stores are a crucial component that stores the created functions. In this structure, the functions approved by the serverless computing system are stored. Clients who request to execute their desired functions will have their requests forwarded to the serverless computing subsystem, which is the API gateway in this case, which will then execute the request. The API gateway structure will call the requested functions from a structure called the functions store, where all the functions are stored [33].

3.2 Blockchain Technology Based on Hyperledger Fabric

Hyperledger Fabric is a blockchain architecture based on a network of distributed nodes that ensures scalability, flexibility, and confidentiality. In contrast to other blockchains, Hyperledger Fabric can run applications that are developed in popular programming languages and can run them on a network of distributed nodes. A blockchain-based system lets developers develop distributed applications in popular programming languages for the first time. A blockchain system based on Hyperledger Fabric technology is a permission-based blockchain architecture that is modular and extensible. A wide range of distributed ledger prototypes, so-called proof of concept projects, and production systems have been developed using Hyperledger Fabric. The applications of this technology include the logistics of trade, the management of contracts, the management of health and safety, the identification, management, and many others [34]. The Hyperledger Fabric architecture allows for the parallel execution of transactions to provide high performance through the endorsement and implementation of each transaction of a subset of peers. In this form of technology, smart contracts must be endorsed by a specified number of peers before they can be considered valid for execution. This process is performed with a customized or default policy for endorsements. This endorsement policy indicates which peers are required to endorse smart contracts. It will only be possible to document the transaction's effect on the ledger state once all peers agree on the deterministic validation method used to implement the transaction. Hyperledger Fabric respects individualized confidentiality assumptions through endorsements of transactions. Furthermore, state updates are delegated to modular components for consensus. It is non-stateful and logically separate from peers that execute transactions and maintain ledgers. A Hyperledger Fabric mechanism as a modular consensus system can customize each deployment's trust assumptions. While achieving peer consensus on a blockchain is possible, the division of responsibilities provides greater flexibility [35]. Consequently, Hyperledger Fabric is a permission-based blockchain platform that supports flexible security assumptions, allowing permission-based blockchains to be built on a scalable and customized platform. Hyperledger Fabric architecture is shown in Fig. 2.

Clients

Clients create a transaction through the Hyperledger Fabric by sending a request to the blockchain system. This transaction can be used to access data or information or insert information into the blockchain database, referred to as the ledger. As a result, the relevant process follows the submission of the clients request. Hence, in this blockchain system, as in other blockchain-based systems, clients are always structural components whose requests are executed as transactions.

Membership Service Provider

A membership service provider is responsible for monitoring and keeping track of all entities' identities, including members and clients (applications), and ensuring they are authenticated and authorized, as well as granting those credentials to verify their identity and authorization. It is essential to know that the Hyperledger Fabric blockchain requires permission. Digital signatures are extensively used in blockchain-based systems Hyperledger Fabric to authenticate all communications between entities (nodes). The membership service component authenticates all blockchain operations at each entity in the network, including authentication, verification, endorsement, and other blockchain operations. As well as registering and managing entities, MSP provides tools for their management [36].

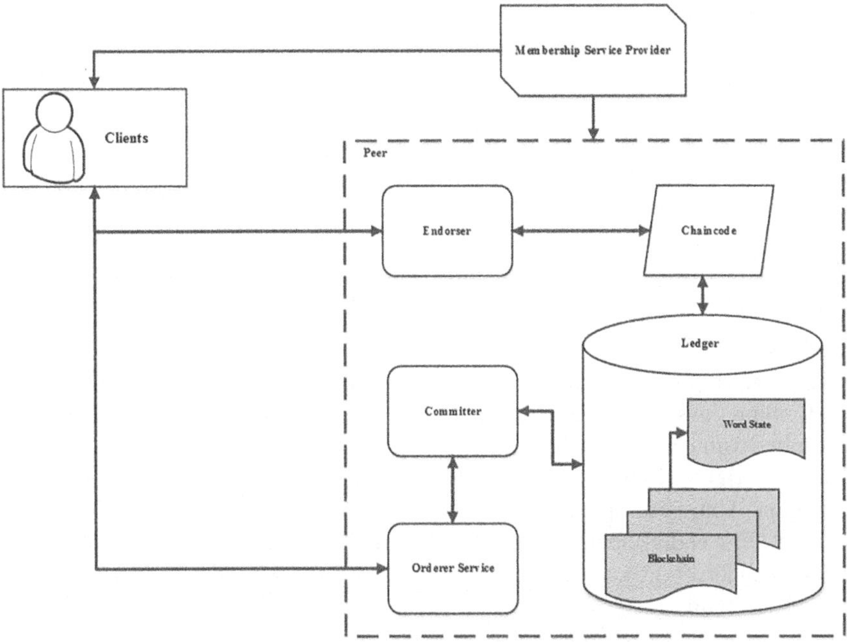

Fig. 2 Hyperledger Fabric architecture

Orderer Service

A Hyperledger Fabric Orderer service requires considering several entities (nodes). This service performed status updates, relationships, and cryptographic signatures to indicate peer endorsements. Status updates, relationships, and cryptographic signatures of peer endorsements are part of the transaction. The ordering service will execute all transactions in the order determined by the ordering service. As orderers are not involved in implementing the applications, they are not aware of their status, so they are not responsible for executing and validating transactions.

Committer

A committer is a Hyperledger Fabric member who is a committed member of the channel. These committed members are responsible for adding valid transactions to the channel ledger. The ledger is also exclusive, and only members with access to these channels can access it. Hence, these members determine which transactions have been validated, then insert those validated transactions into the designated ledger. Peers may not be able to perform both functions in configured environments. However, he may be able to do these things in more open environments [37].

Ledger

A Hyperledger Fabric ledger consists of a blockchain and a world state. Each contains a collection of facts related to a collection of business objects. Blockchains are records of all the modifications that have led to the present world state, and they are a record of transactions, as their name implies. By examining the transactions in the blocks linked to the blockchain, one can observe the history of the developments resulting from the current world state. Once a blockchain data structure has been written down, it cannot be altered again because due to its immutability. On another side, a world state is a database that contains the latest values of a collection of ledger states. Using the world state allows a program to access the latest value of a state directly without having to traverse the whole transaction log. In regard to this, Hyperledger Fabric is flexible in the sense that ledger states are expressed by default as key-value combinations. Since states can be deleted, created, and updated, the world state can modify a lot from time to time.

Endorser
One of the Hyperledger Fabric system's components is an endorser, responsible for performing transaction validation. Validating a policy involves assessing whether it complies with endorsement criteria. In some cases, only designated administrators have the right to change endorsement policies using the system management functions. Untrusted applications cannot select or make changes to endorsement policies. The endorsement policies for Hyperledger Fabric allow the chain code to identify endorsers as a collection of members that must be present for the transaction to be successful.

Chain Code
Chain code is one of the essential elements of a distributed application in Hyperledger Fabric blockchain, or smart contracts, which execute during the execution phase of the distributed application. There is also a term for it called a chain code. To manage the blockchain system, a chain of codes is required. This chain of codes can be generated by anyone, including untrustworthy individuals, and is collectively known as the "system chain codes" [38].

When all of these components are combined, and cooperation can be achieved between them, the result can be the construction of an innovative, highly reliable system that can be used to address a wide variety of security and privacy concerns that may arise in various situations.

4 Security Challenges in Serverless Computing

Serverless computing technology has some disadvantages, distinct advantages, and capabilities, as do other recently introduced technologies. Security concerns have long been a concern for new technologies, including serverless computing, despite their distinctive advantages and capabilities. Hence, security concerns have always been raised about these and other technologies [39]. Microservice architecture is used to implement serverless computing systems, allowing clients to store, invoke, and execute code as a service. Hence, for functions to be deployed, invoked, and executed, the code must be secure and free of malicious code. Otherwise, its implementation could harm both clients and service providers. Serverless computing has caused significant concerns due to inadequate attention to security concerns and some specific subjects discussed previously. The service providers and the individuals using the services are likely to be concerned about each of these difficulties. Several security challenges that may affect the application of this technology will be discussed in this chapter. It is important to note that serverless computing raises significant safety concerns because of these security challenges.

4.1 Masquerade Challenges

The serverless computing system employs a not-so-powerful security mechanism to prevent unauthorized access to services. In this way, the system meets the regulatory requirements so that it complies with them. This tactic is based on the assumption that every client that needs the use of services provided must first be approved of the system's validity, then submit requests for the necessary tasks. The service will be available for use immediately once it has been authorized. This guiding principle serves as the foundation for the overall security strategy. The most significant challenge this mechanism will ever face is when unauthorized clients introduce themselves as authorized clients who have already received authorization previously. The current security mechanisms in serverless computing systems can always encounter this issue. This issue will never disappear from the path of these mechanisms. Once the unauthorized clients have completed this task, they can access the services without the need to utilize an authentication system that is available on serverless computing systems. On the other hand, this event can cause authorized clients not to access the services. The unauthorized clients in question are, therefore, able to utilize the services offered by serverless computing providers and may engage in dishonest or harmful behavior. Through masquerade techniques, unauthorized clients can access serverless computing services through this event, which is feasible due to weak security mechanisms [40].

4.2 Malicious Code Deployment and Execution Challenges

Serverless computing technology can support various software programs directly through a modular code format known as function, as they can adapt, be flexible, and be scalable. One of the benefits of serverless computing systems is that they let developers sell their functions to make money through a market-based framework. By creating such a capability in serverless computing systems, people can quickly prepare the programs they require or—in other words, the functions they require—buy them and run them. This and other unique benefits of serverless computing give programmers more control over the computing environment in which they work. It also lets them create programs in the form of functions that meet the market's needs and sell them on the platforms where they are defined. Hence, many concerns related to this issue include calling and executing functions to achieve desired outcomes and loading those functions into the system. Among the most significant problems this method will face in the future will be the development of destructive code posing as efficient and helpful functions to provide to other clients, either for a fee or for free; this happens when the developer has malicious goals. In light of this, this method will face the most significant security problems. Implementing these functions with malicious code can be bad for clients and providers. Therefore, a severe repercussion for clients and providers is that malicious code is incorporated into the functions [41].

4.3 Privacy Protection Challenges

The existence of technologies that usually reside on one side of service providers and the other side of clients has always raised privacy concerns. In other words, those who provide and use services are usually part of the same issue. Serverless computing providers are responsible for protecting the privacy of their clients. Hence, they must provide safe and risk-free services for their clients. Serverless computing service clients must ensure that their data is safe and secure. They are also protected from any potential abuse, which is the duty of the service providers. Serverless computing uses authentication and verification to make sure clients are who they say they are before they can use the services. Therefore, before utilizing serverless computing services, clients are required to provide this information. Once the security system has access to the data about these people, it needs to complete its authentication procedure correctly. Clients who want to use serverless services must ensure that the security system that protects the serverless service provider's system is safe. Even though these kinds of security systems are almost always attractive to attackers, this is how things have turned out. This is because attackers can get information about clients who use these mechanisms' services if they try to break into the security mechanism system. They can use the information they have to get into the system, which can have severe consequences for both those who use serverless computing services and those who provide them. So, it will be essential to ensure that the security system doesn't pose a security risk [42].

5 Serverless Computing Security with Hyperledger Fabric (SCSHF)

Providers of serverless computing services must verify their clients' identities before delivering their services to them, allowing them to provide services only to clients who can verify their identities. As a result, serverless service providers strive to prevent access to people whose identities have not been verified by defining an authentication system and refrain from offering services to them.

Different viewpoints can be taken into account when explaining this phenomenon. Among the things that can be mentioned is that in the process of using serverless services, clients will be able to load the services they have created in the form of functions into the system and place them in the functions storage structure. A significant issue occurs when people penetrate the authentication mechanism and upload their functions that contain malicious code fragments for their use or even for the use of others. In other cases, an attacker can gain access to the system's services by impersonating other clients, presenting themselves as a client whose identity has already been confirmed, and exploiting this process to abuse these individuals. Therefore, an attacker can abuse clients and providers by utilizing these vulnerabilities. Another challenge in this area is that an attacker can gain access to the system's authentication system in order to learn who has authenticated the system and who can access the system. That means the attacker can access the information of the clients who use the system. Therefore, service providers and clients can experience problems in the event of these issues and other issues related with this field. Because serverless service providers will always worry about the network they have planned for their system being challenged by attackers, this can result in losing clients. Also, clients who use serverless services from providers are concerned about the possibility of their personal information being misused by attackers. Given that they will be using these services, they need reliable security mechanisms to protect their privacy. Using blockchain technology to address the concerns arising in this situation is crucial for reliable security technology, such as blockchain technology, can help to address these concerns. Therefore, using blockchain technology will provide a high level of security for the authentication process and, in turn, will provide clients with a high level of privacy. It is possible to consider Hyperledger Fabric as a suitable blockchain type to address this problem because it uses permission-based blockchain technologies. Hyperledger Fabric allows service providers to customize their blockchain system in accordance with their goals by allowing them to define policies tailored to their goals. This feature, in combination with the powerful capabilities of the blockchain, will help protect the clients information and prevent any abuse or intrusion into the system [43]. This chapter attempts to present a security mechanism called SCSHF (serverless computing security with Hyperledger Fabric) that will enable serverless service providers to use blockchain for the authentication process. This mechanism consists of two separate components, which are:

- Registration at serverless computing security with Hyperledger Fabric (RSC-SHF)
- Authentication in serverless computing security with Hyperledger Fabric (ASC-SHF)

The point that should be noted here is that in this research, an attempt has been made to provide both the client authentication process (ASCSHF) and the client registration process (RSCSHF) for accessing serverless services using the Hyperledger Fabric blockchain. Consequently, we will examine both mechanisms in the following sections. The use of this mechanism is suggested in the scenario of using drones. In this scenario, it is assumed that drones use serverless computing services to process their received information. Regarding this scenario, uncrewed aircraft or drones will be needed to process the information received. Therefore, they must invoke and execute the functions required to process the information they receive so that this event can assist them in reducing energy consumption and ensuring flight continuity. Hence, in this research, we will examine and analyze the challenges faced by serverless service provider systems and drones as clients.

5.1 Registration in Serverless Computing Security with Hyperledger Fabric (RSCSHF)

To use serverless computing services, clients must register at the beginning. Hence, clients interested in using serverless computing services must register by providing their identity and information to the security system designed specifically for serverless computing. Upon authentication, they will be permitted to utilize the serverless computing services. This operation ensures that serverless service providers only provide access to their services to authenticated individuals. As a result of the importance of this issue and the need to provide safe serverless computing services and prevent unauthorized access, the SCSHF mechanism is proposed. Thus, this section examines the mechanism of RSCSHF, which uses blockchain technology to implement the registration process for serverless service clients. According to the scenario presented in this section, drones must first register with the serverless service provider's registration mechanism to access serverless computing services. The registration mechanism presented is an RSCSHF. In this regard, using Fig. 3 as a sequence diagram, this section shows how the RSCSHF mechanism can be implemented and run. This mechanism is designed to prevent unauthorized individuals from accessing serverless computing services and resolve security problems associated with existing registration mechanisms.

In this scenario, it is initially assumed that the drones, here called clients, must completely access serverless computing providers' services. However, they cannot take advantage of this possibility due to the lack of registration. Consequently, in the first stage, clients will attempt to access the system, and at this point, they will send a request to the RSCSHF structure (1). When the RSCSHF structure receives the clients request, it tries to check the clients authentication status in the first step. RSCSHF performs this process to prevent clients from re-registering after previous registration. RSCSHF sends a request to the endorser for this purpose (2). Upon receiving the request from RSCSHF, the endorser sends a request to the chain code and asks the chain code to return the result of the received request by communicating with the ledger through a query (3). As soon as the chain code receives a request from the endorser, it creates a query based on the request and sends it to the ledger to get a response (4). Since the ledger has the nature of a database, it executes the request from China in the form of a query. The requested query will only have the status check nature at this stage. This will mean that this query only requests

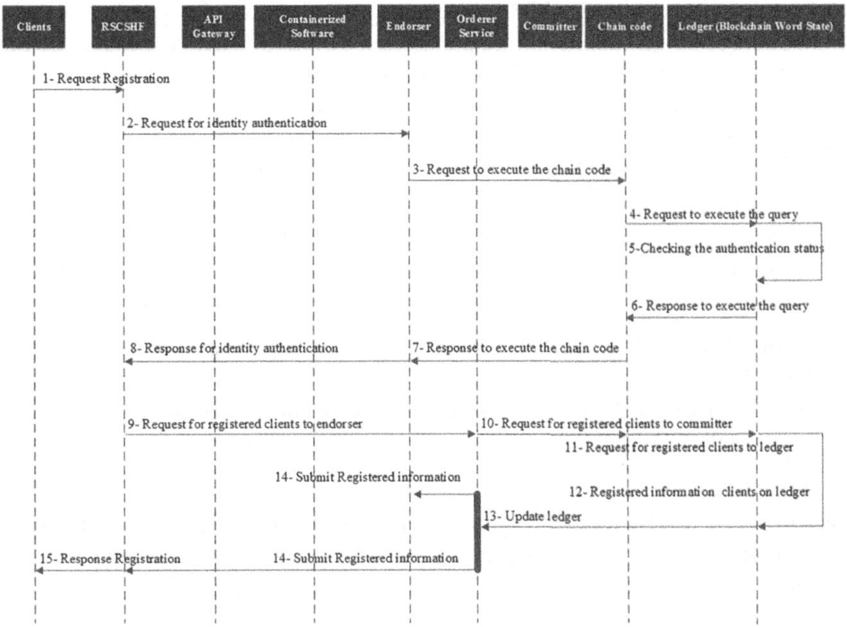

Fig. 3 Registration at serverless computing security with Hyperledger Fabric (RSCSHF) mechanism

information and does not include changes in the ledger that include updating and inserting transactions. Here, it should be noted that the update process is applicable in the word state section, and the primary ledger will only perform the process of inserting transactions, and in this respect, the nature of the ledger cannot be changed (5). After executing the query, the ledger returns the result to the chain code. This result can include two items:

- The clients have already completed the registration process.
- The clients have yet to make the registration process.

This scenario assumes that the clients have never registered before. When the ledger has analyzed the query and executed it, it sends the result of its execution of the chain code by way of a response that this response shows the same as the status of the clients' registration (6). The chain code that receives the result of its request from the ledger tries to send the received result to the endorser in response to the endorser's endorser request when it receives the result from the ledger (7). The endorser also considers the chain code response as a response to the RSCSHF request and attempts to send it to the RSCSHF by creating an appropriate response (8). The RSCSHF mechanism now sends a request to the Orderer service after confirming that the inquiring clients are not registered. It attempts to provide context for execution by adding the received request to the queue, considering the nature of the Orderer service, which is a queue of receiving requests (9). In response to the request from the RSCSHF structure, the Orderer service requests that the committer record the clients' information given by the RSCSHF as a transaction in the ledger (10). After receiving the request from the Orderer service, the committer requests the ledger record the clients' information as a transaction in the ledger (11). Upon receiving the request, the ledger attempts to store related clients' data as a transaction in the ledger blocks (12). After storing the essential clients' information in the transaction format, the ledger attempts to send the result as a response to the Orderer service. This response includes information on the transaction type that caused the ledger to be updated (13). Upon receiving the answer from the ledger, the Orderer service tries to communicate the results to various components, including:

- Sending the results of changing the ledger and adding the clients information to the endorser (14)
- Transmitting to RSCSHF the results of entering client data into the ledger and updating the ledger (14)

When RSCSHF receives a response from the Orderer service, it sends the results back to the clients. By following these instructions, clients can be sure that they have finished the registration process and can now use the serverless computing services (15).

5.2 Authentication in Serverless Computing Security with Hyperledger Fabric (ASCSHF)

As stated in the preceding section, the primary objective of providing such an approach is to determine the most effective method for addressing the security challenges caused by clients' unauthorized access to the services that serverless computing service providers make available. In this regard, this matter is even more significant than the RSCSHF mechanism. This method prevents clients from gaining unauthorized access to serverless computing services. In addition, measures have been taken to prevent individuals from impersonating authorized clients to abuse the services. The ASCSHF mechanism, presented as a proposed mechanism, aims to prevent unauthorized clients from gaining access to the provided services by providing a structure based on the blockchain's unique capabilities. In this regard, using Fig. 4 as a sequence diagram, this section shows how the ASCSHF mechanism can be implemented and run.

In this hypothetical scenario, when clients attempt to use the services provided by serverless computing, the authentication process for those clients should occur immediately. Therefore, they will submit a request to the ASCSHF structure to verify their identity (1). Upon receiving a request, the ASCSHF structure sends a request to the endorser to verify the clients' eligibility to access serverless computing services (2). Immediately after receiving this request, the endorser tries to transmit a request to the chain code to carry out the instructions outlined in the

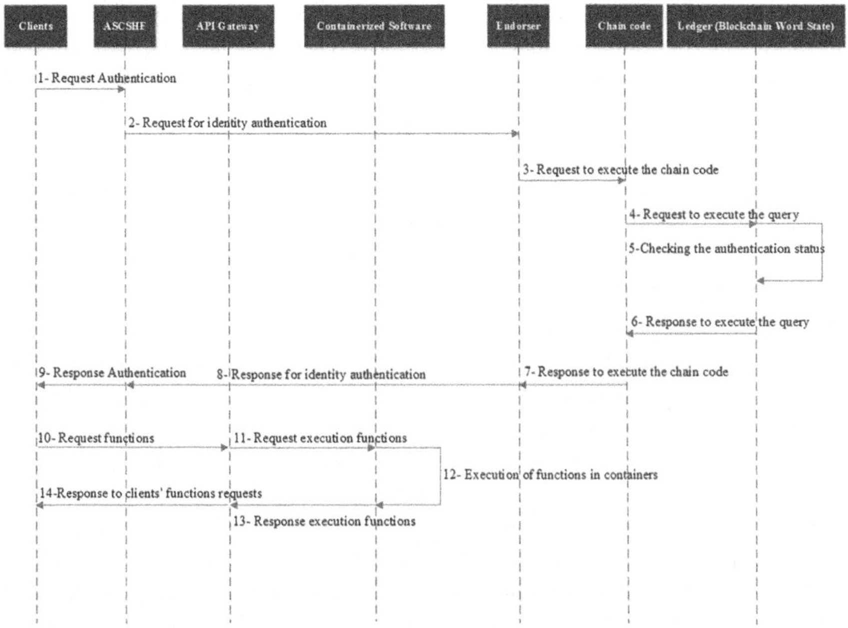

Fig. 4 Authentication in serverless computing security with Hyperledger Fabric (ASCSHF) mechanism

presented request (3). As soon as the chain code receives the request transmitted by the endorser, it attempts to convert the received request into a query and then transmits it to the ledger as a query (4). The ledger structure will attempt to carry out the query submitted after it receives in the form of a query. In the meantime, no changes are made to the blockchain structure. Therefore, after receiving the query, the ledger will check it (5). The following findings are what will come about as a consequence of looking at this query:

- The identity of the requesting client has been validated for security reasons.
- It is impossible to determine the identity of the requesting client has been validated for security reasons.

When the ledger has finished responding to the queries requested, it will make an effort to send the results in the chain code in the form of a response. The ledger will do this right after it has finished responding to the queries that have been requested (6).

When the chain code receives a response from the ledger, it will proceed to create a response to the endorser's request and transmit that response to the endorser. After completing this process, the endorser will receive the response from the chain code (7). As soon as the endorser has obtained the response generated by the chain code, they will make an effort to forward it to ASCSHF to respond to their request. This event will respond to the request submitted earlier in the process (8). Upon receiving the endorsement response from the endorser, the ASCSHF will reply to its clients about the outcome of their endorsement request. Therefore, the ASCSHF will take action to maintain open communications with our clients. Hence, the following types of clients appear to fall into this category:

- The identity of the clients who submitted the request is verified. Therefore, the clients can access the services offered by the serverless computing provider.
- The identity of the clients who submitted the request has not been verified. Therefore, the clients will not be able to access the services offered by the serverless computing provider.

Assuming the identity of the clients who submitted the request has been verified, serverless computing is expected to be able to provide the services to them. This case, therefore, assumes that the clients identity has been verified (9). Upon receiving the result of their request, which confirms that they are confirmed clients, they will strive to invoke the appropriate functions to carry out the required procedures to achieve their goal. As a result, the client can accomplish the objective outlined above. Hence, their requests are submitted to the API gateway (10). When the API gateway receives requests from clients to invoke and execute the functions requested, it will generate a request to fulfill those requests. Therefore, the API gateway will attempt to invoke and execute the functions requested by the clients. Hence, the API gateway transmits requests to containerized software (11). The containerized software attempts to implement the functions requested by the API gateway in the form of containers after receiving the request from the API gateway (12). As soon as the containerized software has executed the functions, it will send the response as a result of the execution of those functions in response to the API gateway's request (13). As a final step, API gateway creates a response and sends it to clients after receiving the response to its request (14). Figure 5 depicts SCSHF (RSCSHF and ASCSHF) mechanism in the context of serverless computing services utilized by drones.

6 Conclusion

Serverless computing has gained increasing acceptance in recent years due to its ability to facilitate the design and delivery of services using code fragments. It is possible to maximize resource use by executing the desired service and minimizing the implementation of unnecessary aspects, which can then be used to reduce resource expenditures. Serverless computing technology is still in its infancy, and like all new technologies, it has some flaws that can make it challenging to use. The lack of effectiveness of this technology in providing high security is one of its most significant concerns. Therefore, serverless service providers require clients to supply identifying information to ensure that only users with verified identities can access their services. Hence, clients must provide information through various verification mechanisms to verify their identities. These security mechanisms must

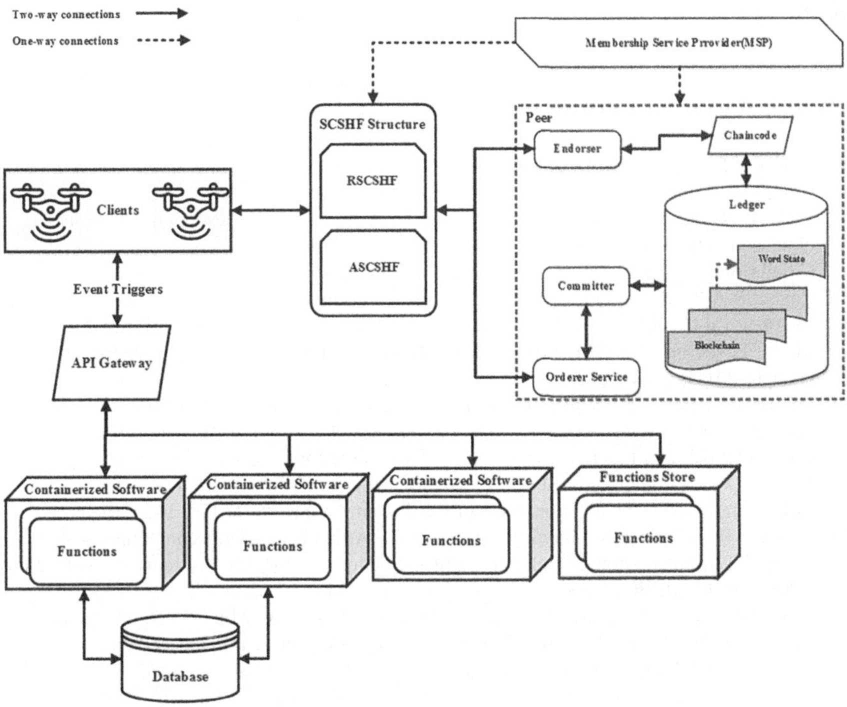

Fig. 5 SCSHF (RSCSHF and ASCSHF) architecture

protect clients' privacy to prevent unauthorized information access. When using sensitive technologies like drones, these security mechanisms can't protect against attacks that try to get access to information or take resources. Using technology such as blockchain, providers can provide a high level of security and significantly reduce security problems in this field. This chapter suggests a new security mechanism known as SCSHF, which integrates two blockchain technologies with serverless computing to provide excellent security features through these technologies. This security mechanism makes it possible to use serverless computing services in sensitive fields, like drones, with adequate security. The proposed SCSHF approach, which has two parts called RSCSHF and ASCSHF, takes advantage of the high level of security that permission blockchain technologies like Hyperledger Fabric offer to make a high-level security system for serverless service providers. This security feature can make clients' private information much safer when using serverless computing services. Consequently, for malicious individuals who seek to target serverless service provisioning systems, this security mechanism can make these assaults difficult or extremely costly, thereby significantly reducing the frequency of these attacks.

References

1. Khmelevsky, Y., Voytenko, V.: Cloud computing infrastructure prototype for university education and research. In: Proceedings of the 15th Western Canadian Conference on Computing Education, pp. 1–5. Association for Computing Machinery, New York (2010)
2. Sowmya, S.K., Deepika, P., Naren, J.: Layers of cloud–IaaS, PaaS and SaaS: a survey. Int. J. Comput. Sci. Inf. Technol. **5**(3), 4477–4480 (2014)
3. Lynn, T., Rosati, P., Lejeune, A., Emeakaroha, V.: A preliminary review of enterprise serverless cloud computing (function-as-a-service) platforms. In: 2017 IEEE International Conference on Cloud Computing Technology and Science (CloudCom), pp. 162–169. IEEE (2017)
4. Rajan, R.A.P.: Serverless architecture-a revolution in cloud computing. In: 2018 Tenth International Conference on Advanced Computing (ICoAC), pp. 88–93. IEEE (2018)
5. Marin, E., Perino, D., Di Pietro, R.: Serverless computing: a security perspective. J. Cloud Comput. **11**(1), 1–12 (2022)
6. Polinsky, I., Datta, P., Bates, A., Enck, W.: SCIFFS: enabling secure third-party security analytics using serverless computing. In: Proceedings of the 26th ACM Symposium on Access Control Models and Technologies, pp. 175–186. Association for Computing Machinery (2021)
7. Hammi, M.T., Hammi, B., Bellot, P., Serhrouchni, A.: Bubbles of Trust: a decentralized blockchain-based authentication system for IoT. Comput. Secur. **78**, 126–142 (2018)
8. Koschel, A., Klassen, S., Jdiya, K., Schaaf, M., Astrova, I.: Cloud computing: serverless. In: 2021 12th International Conference on Information, Intelligence, Systems & Applications (IISA), pp. 1–7. IEEE (2021)
9. Baldini, I., Castro, P., Chang, K., Cheng, P., Fink, S., Ishakian, V., Mitchell, N., Muthusamy, V., Rabbah, R., Slominski, A., Suter, P.: Serverless computing: current trends and open problems. In: Research Advances in Cloud Computing, pp. 1–20. Springer, Singapore (2017)
10. Lloyd, W., Ramesh, S., Chinthalapati, S., Ly, L., Pallickara, S.: Serverless computing: an investigation of factors influencing microservice performance. In: 2018 IEEE International Conference on Cloud Engineering (IC2E), pp. 159–169. IEEE (2018)
11. Pérez, A., Moltó, G., Caballer, M., Calatrava, A.: Serverless computing for container-based architectures. Futur. Gener. Comput. Syst. **83**, 50–59 (2018)
12. Rajan, A.P.: A review on serverless architectures-function as a service (FaaS) in cloud computing. TELKOMNIKA (Telecommunication Computing Electronics and Control). **18**(1), 530–537 (2020)
13. Mohanty, S.K., Premsankar, G., Di Francesco, M.: An evaluation of open source serverless computing frameworks. In: CloudCom, 2018, pp. 115–120. IEEE (2018)
14. Yu, T., Liu, Q., Du, D., Xia, Y., Zang, B., Lu, Z., Yang, P., Qin, C., Chen, H.: Characterizing serverless platforms with serverlessbench. In: Proceedings of the 11th ACM Symposium on Cloud Computing, pp. 30–44. Association for Computing Machinery, New York (2020)
15. Gand, F., Fronza, I., El Ioini, N., Barzegar, H.R., Pahl, C.: Serverless container cluster management for lightweight edge clouds. In: CLOSER, pp. 302–311, Prague, Czech Republic (2020)
16. Castro, P., Isahagian, V., Muthusamy, V., Slominski, A.: Hybrid serverless computing: opportunities and challenges. arXiv preprint arXiv:2208.04213 (2022)
17. Xie, R., Tang, Q., Qiao, S., Zhu, H., Yu, F.R., Huang, T.: When serverless computing meets edge computing: architecture, challenges, and open issues. IEEE Wirel. Commun. **28**(5), 126–133 (2021)
18. Mondal, S.K., Pan, R., Kabir, H.D., Tian, T., Dai, H.N.: Kubernetes in IT administration and serverless computing: an empirical study and research challenges. J. Supercomput. **78**, 2937–2987 (2022)
19. Latif, R.M.A., Farhan, M., Rizwan, O., Hussain, M., Jabbar, S., Khalid, S.: Retail level blockchain transformation for product supply chain using truffle development platform. Clust. Comput. **24**, 1–16 (2021)

20. Peng, L., Feng, W., Yan, Z., Li, Y., Zhou, X., Shimizu, S.: Privacy preservation in permissionless blockchain: a survey. Digit. Commun. Netw. **7**(3), 295–307 (2021)
21. Paul, P., Aithal, P.S., Saavedra, R., Ghosh, S.: Blockchain technology and its types—a short review. Int. J. Appl. Sci. Eng. (IJASE). **9**(2), 189–200 (2021)
22. Liu, P., Chen, A.Y., Huang, Y.N., Han, J.Y., Lai, J.S., Kang, S.C., Wu, T.H., Wen, M.C., Tsai, M.H.: A review of rotorcraft unmanned aerial vehicle (UAV) developments and applications in civil engineering. Smart Struct. Syst. **13**(6), 1065–1094 (2014)
23. Vergouw, B., Nagel, H., Bondt, G., Custers, B.: Drone technology: types, payloads, applications, frequency spectrum issues and future developments. In: The Future of Drone Use: Opportunities and Threats from Ethical and Legal Perspectives, pp. 21–45. T.M.C. Asser Press, The Hague (2016)
24. Gong, J., Li, D., Yan, J., Hu, H., Kong, D.: Comparison of radar signatures from a hybrid VTOL fixed-wing drone and quad-rotor drone. Drones. **6**(5), 110 (2022)
25. Elijah, T., Jamisola, R.S., Tjiparuro, Z., Namoshe, M.: A review on control and maneuvering of cooperative fixed-wing drones. Int. J. Dyn. Contr. **9**, 1332–1349 (2021)
26. Velliangiri, S., Karthikeyan, P.: Blockchain technology: challenges and security issues in consensus algorithm. In: 2020 International Conference on Computer Communication and Informatics (ICCCI), pp. 1–8. IEEE (2020)
27. O'Meara, W., Lennon, R.G.: Serverless computing security: protecting application logic. In: 2020 31st Irish Signals and Systems Conference (ISSC), pp. 1–5. IEEE (2020)
28. Bardsley, D., Ryan, L., Howard, J.: Serverless performance and optimization strategies. In: 2018 IEEE International Conference on Smart Cloud (SmartCloud), pp. 19–26. IEEE (2018)
29. Spillner, J.: Quantitative analysis of cloud function evolution in the AWS serverless application repository. arXiv preprint arXiv:1905.04800 (2019)
30. Alpernas, K., Flanagan, C., Fouladi, S., Ryzhyk, L., Sagiv, M., Schmitz, T., Winstein, K.: Secure serverless computing using dynamic information flow control. arXiv preprint arXiv:1802.08984 (2018)
31. Pérez, A., Moltó, G., Caballer, M., Calatrava, A.: A programming model and middleware for high throughput serverless computing applications. In: Proceedings of the 34th ACM/SIGAPP Symposium on Applied Computing, pp. 106–113. Association for Computing Machinery, New York (2019)
32. Kaviani, N., Kalinin, D., Maximilien, M.: Towards serverless as commodity: a case of Knative. In: Proceedings of the 5th International Workshop on Serverless Computing, pp. 13–18. Association for Computing Machinery, New York (2019)
33. Sampé, J., Sánchez-Artigas, M., García-López, P., París, G.: Data-driven serverless functions for object storage. In: Proceedings of the 18th ACM/IFIP/USENIX Middleware Conference, pp. 121–133. ACM (2017)
34. Bryatov, S.R., Borodinov, A.A.: Blockchain technology in the pharmaceutical supply chain: Researching a business model based on Hyperledger Fabric. In: Proceedings of the International Conference on Information Technology and Nanotechnology (ITNT), vol. 10, pp. 1613–0073, Samara, Russia (2019)
35. Aggarwal, S., Kumar, N.: Hyperledger. In: Advances in Computers, vol. 121, pp. 323–343. Elsevier (2021)
36. Brotsis, S., Kolokotronis, N., Limniotis, K., Bendiab, G., Shiaeles, S.: On the security and privacy of Hyperledger Fabric: challenges and open issues. In: 2020 IEEE World Congress on Services (SERVICES), pp. 197–204. IEEE (2020)
37. Ren, L., Zhou, H., Hang, X., Yang, B., Su, L.: Research on performance optimization and application in smart home for Hyperledger Fabric. Sensors. **22**(9), 3222 (2022)
38. Androulaki, E., Cachin, C., De Caro, A., Kind, A., Osborne, M.: Cryptography and protocols in Hyperledger Fabric. In: Real-World Cryptography Conference, pp. 12–14 (2017)
39. Ghobaei-Arani, M., Ghorbian, M.: Scheduling mechanisms in serverless computing. In: Serverless Computing: Principles and Paradigms, pp. 243–273. Springer International Publishing, Cham (2023)

40. Mtita, C., Laurent, M., Sauveron, D., Akram, R.N., Markantonakis, K., Chaumette, S.: Serverless protocols for inventory and tracking with a UAV. In: 2017 IEEE/AIAA 36th Digital Avionics Systems Conference (DASC), pp. 1–11. IEEE (2017)
41. Qiang, W., Dong, Z., Jin, H.: Se-lambda: securing privacy-sensitive serverless applications using SGX enclave. In: Security and Privacy in Communication Networks: 14th International Conference, SecureComm 2018, Singapore, 8–10 Aug 2018, Proceedings, Part I, pp. 451–470. Springer International Publishing, Singapore (2018)
42. Shafiei, H., Khonsari, A., Mousavi, P.: Serverless computing: a survey of opportunities, challenges, and applications. ACM Comput. Surv. **54**(11s), 1–32 (2022)
43. Ghorbian, M., Ghobaei-Arani, M.: A blockchain-enabled serverless approach for IoT healthcare applications. In: Serverless Computing: Principles and Paradigms, pp. 193–218. Springer International Publishing, Cham (2023)

Utilizing Blockchain for Safeguarding IoT-Based Robotic Networks from Spoofing Attacks

Tauhidul Alam and Sajedul Talukder

Abstract In the dynamic realm of IoT-based robotic networks, autonomous robots, with their intricate integration of sensors and data-driven functionalities, become prime targets for cyber threats. Their heightened dependence on sensor data for path planning and control makes them particularly vulnerable. Within these multi-robot ecosystems, the process of achieving consensus is vital for ensuring coordinated communication. However, this consensus mechanism can become a weak point, susceptible to adversarial intrusions such as spoofing attacks. In this chapter, we introduce a novel approach that leverages blockchain to facilitate path planning coordination among a network of robots. Furthermore, we present a consensus algorithm for detecting compromised robots by analyzing blockchain data transferred from the robots. Through simulation results, we demonstrate that the proposed approach enhances the resilience of a robot network by rapidly identifying spoofed client robots or compromised servers. We conclude with a comprehensive discourse on the multifaceted benefits of deploying blockchain as a defensive measure in these networks while also shedding light on the emergent challenges and their feasible countermeasures.

Keywords Robotic networks · Spoofing attack · Blockchain · Communication

1 Introduction

The rapid evolution of technology has ushered in a new era where the Internet of Things (IoT) and robotics converge, crafting intricate networks that promise

T. Alam
Department of Computer Science, Louisiana State University Shreveport, Shreveport, LA, USA
e-mail: talam@lsus.edu

S. Talukder (✉)
School of Computing, Southern Illinois University, Carbondale, IL, USA
e-mail: sajedul.talukder@siu.edu

N. El Madhoun et al. (eds.), *Building Cybersecurity Applications with Blockchain and Smart Contracts*, Signals and Communication Technology,
https://doi.org/10.1007/978-3-031-50733-5_4

transformative changes across industries. These IoT-based robotic networks, symbolizing the pinnacle of automation and interconnectivity, have found applications spanning from smart manufacturing and healthcare to advanced urban infrastructure and beyond. Robot networks have made a notable impact in several applications, including drone delivery, infrastructure inspections, disaster information gathering, agriculture precision, border and area surveillance, and search and rescue operations. An example application of robot networks is Wing's drones, which have received certification from the Federal Aviation Administration (FAA) in the USA. These drones efficiently deliver small packages, such as food, medicine, and household items, directly to homes in minutes following flight paths [43]. To achieve seamless operation, these robots within the network use wireless communication sensors for their path planning, coordination, control, and collision avoidance. However, the use of wireless communication sensors exposes them to potential security risks. Malicious agents could exploit this vulnerability by jamming or intercepting these wireless sensors, thereby gaining unauthorized access to the network and making the robots susceptible to cyberattacks and malicious traffic. In particular, a robot network can be disrupted by the spoofing attack, also referred to as a Sybil attack [5]. In such attacks, adversaries can forge multiple spurious identities or impersonate several existing client robots within the network [29] after having complete control over critical sensors like GPS [35] or optical flow sensors [3].

The motivation for our work stems from the challenge of protecting a robot network against spoofing attacks. The spoofing attack involves an adversary luring a set of client robots away from their intended service robot, as depicted in Fig. 1. Our focus is on a specific scenario, where a group of aerial delivery vehicles operates in a robot network, transporting packages from a central fulfillment or distribution station to designated customer (goal) locations. These delivery vehicles are under the control of a server robot. However, the network faces a significant threat as the adversary seeks to gain control over multiple delivery vehicles by spoofing their customer locations or compromising the server robot. This spoofing attack is easy to carry out but difficult to prevent in multi-robot settings. Consequently, our research is centered on the challenge of safeguarding the network by detecting this attack and thwarting this malicious activity.

Our work is closely related to the development of spoof-resilient solutions to multi-robot networks using information extracted from Wi-Fi communication signals for detecting spoofed client robots [14] and providing a consensus algorithm with bounded performance guarantees [13]. Consensus methods in mobile and distributed networks have also been explored, involving the use of transmitted values [33, 34] to remove adversarial agents from the consensus and the use of exchanged keys [18] or tokens [20] to secure networks with different initial topologies [42]. However, these consensus methods are prone to failures when robots within the network malfunction or transmit incorrect messages. Additionally, malicious agents can generate a number of false identities in a robot network that utilizes wireless signals for security instead of using a trusted system within the network. In contrast to prior works, our approach adopts the use of trusted blockchain technology to ensure resilient coordination and communication among

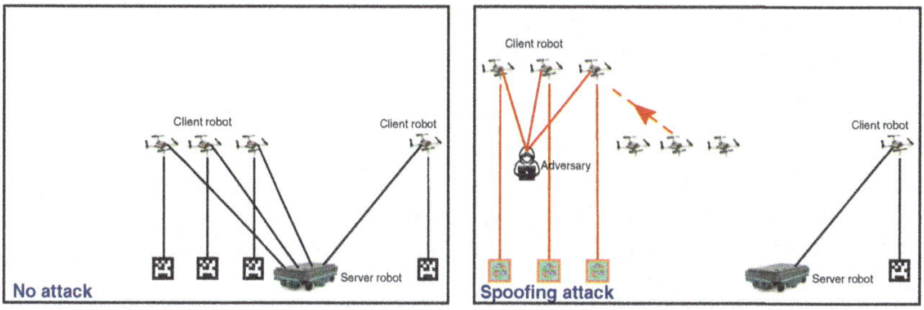

Fig. 1 Spoofing attack. A server robot controls pre-computed flight paths to client robots for their intended customer locations when no attack is present. In the spoofing attack, an adversary spoofs many client robots by diverting them from their original paths and redirecting them toward pseudo-customer locations that the adversary controls

robots within the network, even in the presence of malfunctioning and malicious robots (Byzantine robots).

The potential benefits of using blockchain technology to tackle security issues in swarm robotic systems are discussed in [10]. The Nakamoto's white paper [22] introduces blockchain technology as a trusted database of encrypted and linked data transactions with timestamps stored by participating agents of a peer-to-peer network. There are limited approaches that currently explore the utilization of blockchain technology in robotic security systems. One such approach is a blockchain-based collective decision-making approach [37, 39] for managing Byzantine robots in homogeneous robot swarms. Additionally, blockchain has been utilized in heterogeneous robot swarms to facilitate collaboration [27]. Existing consensus algorithms are compared with the blockchain consensus algorithm in [38]. The pivotal advantages of using blockchain are the immutability of transactions, decentralized consensus, fault tolerance, and so on. In this context, this work leverages a permissioned or private blockchain, where a centralized entity has control over its participants. This approach is employed to detect spoofed client robots or compromised servers by validating them through a committee of robots within the network. Leveraging blockchain technology for spoof-resilient robot networks can provide several benefits in terms of security, trust, and transparency. Blockchain, as a decentralized and immutable ledger, can help prevent spoofing attacks and significantly enhance the overall integrity of robot networks. Here are some concrete advantages of blockchain technology that can be utilized to make robot networks secure, resilient, and trustworthy, effectively combating spoofing attacks and other potential threats.

Identity Management Blockchain can be used to secure and tamper-proof identity management of robots within the network. Each robot has a unique digital identity stored on the blockchain, and any attempt to tamper with the identity becomes detectable and preventable. This guarantees that only authorized robots can participate in exchanging data within the network. Additionally, employing common blockchain techniques like hashing and time-stamping enhances the security and validation of data shared among the robots.

Authentication and Access Control Blockchain-based authentication mechanisms can be employed to ensure that only legitimate robots can access network resources. Through the use of smart contracts, access control policies are defined, and robots can authenticate themselves using cryptographic keys stored on the blockchain. This robust approach prevents untrusted robots from gaining unauthorized access to network resources.

Data Integrity and Auditing Blockchain provides an immutable and transparent record of all transactions and activities within the network. Storing robot-generated data on the blockchain makes it becomes exceedingly difficult for malicious actors to manipulate the data. This significantly enhances the integrity of the data collected by robots and enables easy auditing and verification of its authenticity.

Consensus and Trust Blockchain networks rely on consensus mechanisms to validate and agree upon the state of the network. Leveraging consensus algorithms, as discussed in this chapter, help mitigate spoofing attacks. Any attempt to tamper with the data or introduce fraudulent transactions is promptly rejected by the network. Consequently, this fosters a higher level of trust among the participants in the robot network.

Supply Chain and Traceability If the robot network involves supply chain operations, blockchain can be utilized to track and verify the movement of goods or components. Each transaction or transfer can be recorded on the blockchain, ensuring the authenticity and provenance of the items. This helps prevent masquerading or counterfeiting of goods within the supply chain.

Decentralization and Resilience The decentralized nature of blockchain technology makes it highly resilient to single points of failure and reduces the risk of spoofing attacks. The distributed ledger ensures that no single entity has control over the network, making it harder for malicious actors to manipulate or change the system. This decentralized approach enhances the overall security and integrity of the robot network, making it more resistant to potential attacks and unauthorized changes.

Contributions This chapter makes the following contributions:

- A consensus algorithm that utilizes a committee comprising a subset of robots to detect compromised robots within the network by analyzing transferred data transactions on the blockchain.
- A simulation study that validates the performance of our approach for different types of compromised robots in dealing with the spoofing attack.

The structure of this chapter is laid out in a systematic manner to ensure clarity and coherence in presentation. We commence with an exploration of the related work, setting the foundational understanding for our discourse. Following this, we lay out several essential notations tailored for our robot network context and proceed to frame our focal problem in Sect. 3. Thereafter, our approach dedicated to detecting spoofing attacks within the network is delineated in Sect. 4. Empirical evidence from our method's implementation is presented in Sect. 5. In the subsequent section, Sect. 6, we shed light on both the merits and the challenges of harnessing blockchain technology for bolstering security in robotic networks while also suggesting potential countermeasures. The chapter culminates with a concluding synthesis and insights into prospective research avenues in Sect. 7.

It is worth noting that an earlier version of this chapter appeared in [1]. This chapter adapts and extends that work with discussions by exploring the potential of blockchain technology within the realm of robotic networks.

2 Related Work

The rapid proliferation of Internet of Things (IoT) devices offers a myriad of opportunities, but it equally presents a plethora of challenges concerning security, privacy, and trust [7, 30, 36]. As these IoT devices start to encompass autonomous robots, these cyber-physical systems have grown to become prime targets for adversarial activities. One notable vulnerability is their high dependence on sensor data, essential for functions like path planning and control [4, 15]. In multi-robot systems, the act of achieving consensus, paramount for effective coordination, can be manipulated to introduce incoherent decisions, as revealed by Fischer et al.'s work on the impossibility of distributed consensus [12].

IoT-robotic convergence has unlocked numerous opportunities, pushing the boundaries of what can be achieved in automation [19]. While these systems offer great promise, their rising complexity also makes them increasingly vulnerable to cyberattacks [8]. Consensus in robotic networks, particularly in IoT-driven environments, has been a subject of extensive research over the past decade [16]. The importance of consensus in multi-agent systems, ensuring that robots achieve a unified understanding of their environment and tasks, is critical for collaborative tasks [21]. However, this need for consensus creates inherent vulnerabilities, especially when external malicious entities attempt to disrupt it.

Spoofing attacks pose one of the most severe risks in these contexts, leading to robots receiving misleading information [41]. Recent research highlights the susceptibility of sensor data in IoT-based robotic systems, emphasizing the importance of fortifying these data streams [9].

One of the most innovative solutions that have emerged to enhance security in IoT-robotic systems is the integration of blockchain technology [2]. Blockchain's decentralized architecture offers robustness against many conventional attacks, making it a compelling choice for these applications. Recent developments, such as the work by Nguyen et al. [23], showcase blockchain's potential in access control for IoT devices. Meanwhile, the intersection of smart homes and blockchain, as explored by Zhang et al. [44], provides insights into practical deployments of this technology.

There's been an exponential growth in literature emphasizing the harmonious blend of blockchain and IoT [17, 25]. Critical aspects, including blockchain's consensus mechanisms tailored for IoT, have been discussed in-depth in recent studies like that of Zhao et al. [45]. Additionally, the novel framework proposed by Rahman et al. [28], focusing on secure data exchanges in industrial IoT using blockchain, underscores the technology's growing influence in diverse sectors.

In light of the aforementioned literature, it becomes evident that while strides have been made in understanding and mitigating vulnerabilities in robotic networks, there remains a lacuna in holistic solutions that synergistically combine the strengths of blockchain. Our chapter aims to bridge this gap, offering a fresh perspective on harnessing blockchain for IoT-based robotic networks, particularly against spoofing attacks.

3 Preliminaries

The integration of IoT capabilities with robotic systems has given rise to the concept of a server robot and its many client robots within an IoT-based network. This architecture, while enhancing collaborative functionalities, also brings forth specific challenges, particularly in the realm of cybersecurity. A server robot acts as the central control unit or the primary node in this networked system. Equipped with advanced processing capabilities, storage, and communication faculties, the server robot manages, coordinates, and relays instructions to its client robots. These client robots, often specialized for various tasks, are integrated with IoT sensors and devices that allow them to gather data and perform actions based on the directives from the server robot. The synergy between the server robot and its client robots paves the way for more complex operations, adaptive responses, and efficient task distribution.

However, the interdependence of the server and client robots within the network underscores a significant vulnerability: the risk of spoofing attacks. In a spoofing scenario, malicious entities imitate legitimate devices, be it the server robot or any of the client robots, within the network. Given the hierarchical nature of this system, a compromised server robot or even a single deceptive client robot can lead to a cascading impact on the entire network's operations. Unauthorized data access, disruption of coordinated actions, or even the usurping of control from genuine devices are potential repercussions of a successful spoofing attack.

Addressing such threats necessitates a multi-pronged cybersecurity strategy. It's imperative to authenticate the identity and integrity of each robot within the network continuously. Employing cryptographic techniques, anomaly detection systems, and robust authentication protocols can curtail the likelihood of spoofing. Particularly, a two-way authentication between the server robot and its client counterparts can add an additional layer of security.

3.1 Attack Model

We examine a robot network setting in which m client delivery robots (drones) $\mathcal{D} = \{D_1, \ldots, D_m\}$ obtain computed flight paths $\mathcal{T} = \{\tau_1, \ldots, \tau_m\}$ for their product delivery from a server robot S located at a central distribution center or service station. These client delivery robots are identified by a set of identification keys which are denoted as $\mathcal{I} = \{i_1, \ldots, i_m\}$. They communicate their path information (location, velocity, time, distance, etc.) with the server robot through the identification keys $\mathcal{I}$. Let $P = \{p_1, \ldots, p_m\}$ denote the client delivery robots' locations in $\mathbb{R}^3$. Let $V = \{v_1, \ldots, v_m\}$ denote the client delivery robots' velocities in $\mathbb{R}$. Let $G = \{g_1, \ldots, g_m\}$ denote the client delivery robots' goal or customer locations in $\mathbb{R}^2$. Let p_s be the location of S in $\mathbb{R}^2$. Let a flight path of a client delivery robot D_k, where $k \in \{1, \ldots, m\}$, be $\tau_k : [0, t] \to \mathbb{R}^3$ such that $\tau_k(0) = p_s$ and $\tau_k(t) = g_k$ for a finite time interval $[0, t]$. We consider in this setting that either a subset of client delivery robots denoted by $\mathcal{A}$, where $\mathcal{A} \subset \mathcal{D}$, can be spoofed, or the server S is compromised by adversaries. It is assumed that an adversary sends various messages over the network with identification keys to make client delivery robots spoofed by taking control over their GPS sensors and that the knowledge of which client delivery robots are spoofed is unknown [13]. Furthermore, all the client delivery robots are considered to be spoofed when the server robot is compromised. In this context, we formulate the following problem of interest.

Problem (Detecting Spoofed Robots) Given m client delivery robots, the server robot S in a network and their pre-computed paths $\mathcal{T}$ detect $\mathcal{A}$ spoofed client delivery robots or the compromised server S in the case of the spoofing attack.

4 Methodology

This section details our methodology based on the blockchain leveraged consensus algorithm for detecting the spoofing attack in a robot network.

In our approach, we first construct a server robot, and then it establishes a network with m client delivery robots. The server robot provides the identification keys $\mathcal{I}$ to client delivery robots for communication and sends their pre-computed paths $\mathcal{T}$ toward their goal locations G.

In our next step, we employ a private blockchain on the server robot in order to keep track of transferred data over the network. Client delivery robots communicate data related to their locations, velocities, distances, and time periodically with the server robot, and the server robot incorporates them into the transferred blockchain data. The transferred blockchain data is defined as B. We assume that client delivery robots act honestly in communicating their data.

Afterward, we develop a consensus Algorithm 1 in the robot network for detecting the spoofing attack. To achieve this, we devise a verification committee with the server robot and a subset of random client delivery robots. The server

Algorithm 1: CONSENSUS $(\mathcal{Q}, \mathcal{T}, \mathcal{I}, B, w_c)$

Input: $\mathcal{Q}, \mathcal{T}, \mathcal{I}, B, w_c$ – Consensus trigger, Client delivery robots' paths, Set of identification keys for client delivery robots, blockchain data, Weight of each committee member

Output: $\mathcal{L}$ – Set of votes for client delivery robots

for $k = 1$ **to** $|\mathcal{I}|$ **do**
  $t \leftarrow$ PATHDEVIATION(B, i_k, τ_k)
  if $t < \pi$ **then**
    $\mathcal{L} \leftarrow \mathcal{L} \cup \{-w_c\}$
  else
    $\mathcal{L} \leftarrow \mathcal{L} \cup \{w_c\}$
return $\mathcal{L}$

robot alone can also detect spoofed client delivery robots but cannot detect itself being compromised. Let $\mathcal{C} \subset \mathcal{D} \setminus \mathcal{A}$ be the subset of client delivery robots in the verification committee. It is considered that n client committee members, where $|\mathcal{C}| = n$ and $n < m$, can access the transferred blockchain data B. We also consider that both the server and the client committee members are not compromised at the same time.

The verification committee members can vote in a weighted manner for detecting spoofed client delivery robots or a compromised server. Let w_c be the weight for each committee member, including the server robot. Let w_s be the weight for the server robot's vote and w_{cc} be the weight for each client committee robot's vote. The client committee robot's weight for voting is calculated as $w_{cc} = (1 - w_s)/n$. The weights for all committee members can be represented as a $(n+1)$-dimensional vector as follows:

$$w = (w_1, \ldots, w_{n+1}) = (w_s, w_1, \ldots, w_n).$$

The elements of w should satisfy two conditions: (1) $w_c > 0$ for all $c \in \{1, \ldots, n+1\}$ and (2)

$$\sum_{c=1}^{n+1} w_c = w_s + \sum_{cc \in \mathcal{C}} w_{cc} = 1.$$

Let $\mathcal{Q}$ be the consensus trigger by the first committee member to start the consensus process. Let $\mathcal{L} = \{l_1, \ldots, l_m\}$ be the set of each committee member's votes for client delivery robots. Let π be the threshold value for the path deviation of a client delivery robot. In Algorithm 1, we calculate the set of votes for client delivery robots to determine one or more spoofed client delivery robots. We apply the consensus Algorithm 1 for each verification committee member, including the server robot. For each client delivery robot with an identification key i_k, we check its path deviation from the provided path τ_k by the server robot. Since we consider that a client delivery robot's GPS is spoofed, we make use of its locations data that are stored on the blockchain to find the path deviation. Thus, the PATHDEVIATION function takes the input of blockchain data B, identification key i_k, and path τ_k with the location of a client delivery robot's original destination. The function iterates through B to compare the client delivery robot's current location to its goal location to determine if the client delivery robot is getting closer to its intended destination. Once determined, the function returns a value t between 0 and 1. If it is less than the threshold value π, the algorithm determines that the client delivery robot is deviating from its intended path. Then, we add $-w_c$ to the set of votes for that client delivery robot. Otherwise, we add w_c to the set for the same client delivery robot.

In our final step, we investigate two spoofing attack scenarios: (1) client delivery robots are spoofed, and (2) the server is compromised. The verification committee validates these attack scenarios utilizing the transferred blockchain data B by detecting the compromised server or spoofed client delivery robots. In both scenarios, one of the client committee members begins the consensus by voting for each client delivery robot and alerting the other committee members, including the server, to initiate their voting. Therefore, the rest of the committee members also provide their votes for each client delivery robot. Once a consensus is reached by all the committee members, the total votes are combined to a tally for each client delivery robot. If the vote tally for a client delivery robot is less than zero, the associated client delivery robot is detected as spoofed.

For the first scenario, a set of client delivery robots $\mathcal{D}$ is launched with n client committee members (robots) by the server. A subset of client delivery robots $\mathcal{A}$ is spoofed and changes directions mid-way to their intended destinations. The committee members detect a spoofed client delivery robot through their consensus algorithm after it moves away from its intended destination. It is important to mention for this scenario that the server robot itself can also detect a spoofed client delivery robot without the verification of other client committee members as both the server and client committee members can utilize the transferred blockchain data B for verification.

For the second scenario, the server is no longer considered as a committee member. The n client committee members are relied upon to complete the consensus using their accessible communicated blockchain data B. Once the consensus is initiated and completed, the votes are tallied and checked. All the client delivery robots will be detected as spoofed because the compromised server automatically spoofs non-committee client members. Once all non-committee client members are detected as spoofed, the server will be detected as compromised.

5 Experimental Results

In this section, we present the results from the implementation of our approach.

Initially, we implemented a robot network through server-client socket programming in Python 3 with the simulation of port numbers and identification keys for

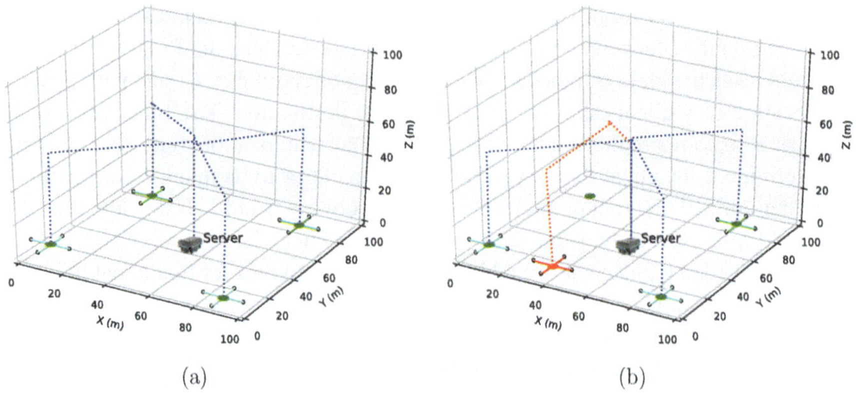

Fig. 2 Client delivery drone spoofing scenario. Flight paths of two client committee drones (depicted in cyan) toward their desired goal locations (green circles) and one non-committee drone (depicted in green) toward its desired goal location and another (depicted in red) toward its unintended goal location (red circle) start from the server robot's location. Between two non-committee drones, the red non-committee drone is spoofed and detected

robots. The server robot provided flight paths to client delivery robots (drones). Then, we simulated our own private blockchain in this network setting using Python. Client delivery drones followed the provided flight paths which were simulated by taking advantage of a Python Robotics tool [26]. While client delivery drones followed the simulated paths, they communicated their locations, velocities, covered distances, and time with the server robot. These communicated data were transferred over the network through our implemented blockchain.

We also implemented our consensus Algorithm 1 in simulation. In our implementation, we accounted for $n = 2$ random client delivery drones and the server robot for devising the verification committee. These committee members employed transferred blockchain data for validation of non-committee client delivery drones' path deviation using their locations along their flight paths. For the voting process of the verification committee members, the weights we utilized for the server robot and the client delivery drone are $w_s = 0.4$ and $w_{cc} = 0.3$, respectively. The threshold value $\pi = 0.9$ was used for finding the path deviation of a client delivery drone. Finally, the verification committee members recorded their votes for detecting spoofed client delivery drones.

Figure 2a delineates a drone delivery network setting, where $m = 4$ client delivery drones were launched from the server robot's location toward their green goal locations with $n = 2$ client committee drones marked in cyan and the remaining client delivery drones marked in green. Figure 2b presents the first spoofing attack scenario, where one of the non-committee client drones was attacked during its flight path execution and moved away from its intended goal location. In our implementation, the verification committee members detected the client delivery drone as spoofed based on their votes and turned it red, including its path and spoofed goal location. Figure 3 illustrates the variations of distinct velocities (roll,

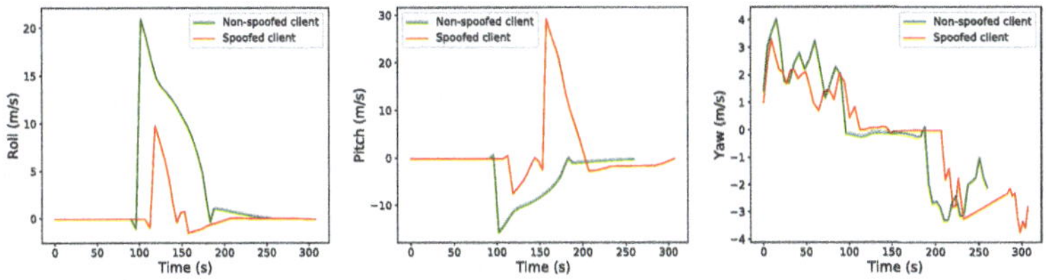

Fig. 3 Comparison of velocities and time for a non-spoofed client delivery drone and its spoofed counterpart after detection

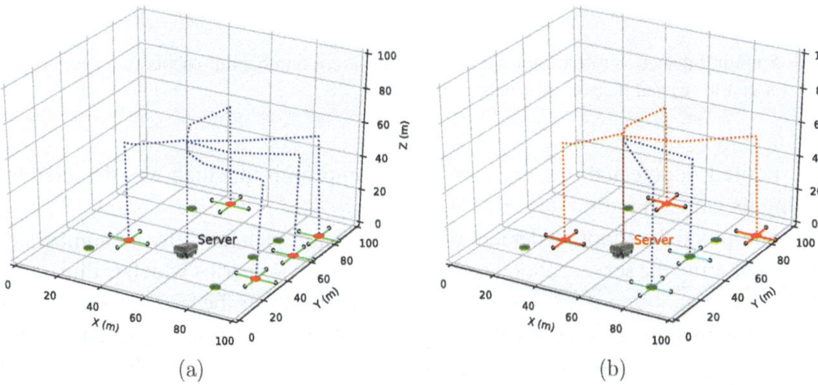

Fig. 4 Server robot compromising scenario. (**a**) Flight paths of five client delivery drones (depicted in green) start from the server robot's location toward their spoofed goal locations (red circles) where the server robot is compromised and undetected. (**b**) Flight paths of two client committee drones (depicted in cyan) toward their desired goal locations (green circles) and three non-committee drones (depicted in red) toward their spoofed goal locations (red circles) start from the server robot's location where it is detected that the server robot is compromised

yaw, pitch) with respect to time for a non-spoofed client delivery drone and its detected spoofed counterpart. These results indicate the inconsistencies in velocities between an actual flight path and a deviated flight path of a client delivery drone when it is attacked.

Figure 4a shows a spoofing attack scenario where $m = 5$ drones were launched; however, the server was compromised and could not run its own consensus nor could any of the five drones launch their own consensus. Since the server was compromised, it redirected all non-committee drones to spoofed goal locations. Since no committee drones were present, this resulted in no drones being detected as spoofed. Figure 4b demonstrates the results of the second spoofing attack scenario, where five client delivery drones were launched but later the server was compromised. In this case, we converted $n = 2$ client delivery drones into committee members. As a result, the committee members were able to detect the remaining client delivery drones as spoofed. Since all non-committee drones

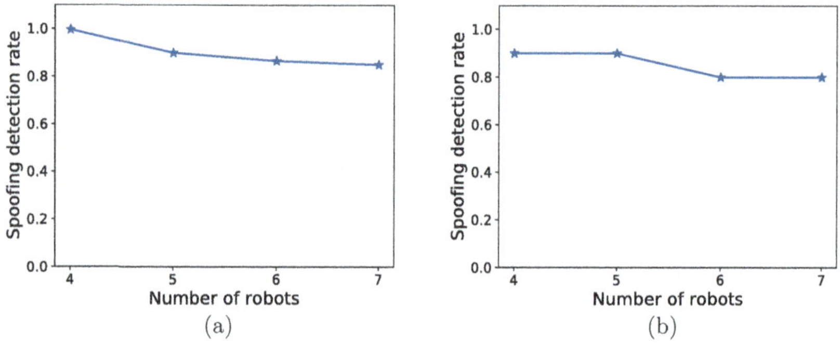

Fig. 5 Spoofing detection rate of our approach for different numbers of robots in a network for the first scenario (**a**) and the second scenario (**b**)

were spoofed, it detected that the server was compromised which was depicted by coloring its name to red.

We computed the spoofing detection rate of our approach for different numbers of robots, including the server robot, in a network for both scenarios which is illustrated in Fig. 5. The spoofing detection rate was computed from the average of 10 runs of our implementation for each number of robots for both scenarios. This result shows that our detection rate is significant but decreases slightly with the increase in the number of robots. The reason for this small detection rate decline is that the client committee members sometimes complete their flights or do not even start their flights, while some non-committee client members are spoofed. This problem can be overcome by dynamically assigning client committee members that are on their flights to the verification committee.

6 Discussions: Challenges and Mitigations

As we delve deeper into the intricate realm of IoT-based robotic networks, especially within the architecture of server robots and their multiple client counterparts, it becomes evident that the potential benefits are counterbalanced by a set of challenges. These challenges, rooted in both the technological intricacies and the operational dynamics of such networks, can potentially impede their efficiency, reliability, and security. However, with every challenge comes an opportunity for mitigation, a solution that can harness the network's full potential while addressing its vulnerabilities. In the ensuing sections, we will explore the most pressing challenges faced by these networks and the strategies that can be employed to navigate and counteract them.

Computational Overhead Blockchains can suffer from computational overhead when there are too many robots on a network. The validation process in a blockchain requires significant computational and memory resources. In a robot network with limited computational capabilities, this can lead to communication delays and hinder the overall performance of the system.

Lack of Scalability Unlike centralized systems, blockchains face challenges in handling a large number of transactions and a growing network size. As the number of robots increases, the processing speed of transactions can slow down, leading to higher costs. This can be problematic for real-time solutions such as robot networks that demand fast speeds and high throughput.

Storage Requirements The distributed nature of blockchain means that all information is shared and stored across different robots within a network, which can lead to high storage requirements over time. Managing and scaling this storage can be challenging, especially as the network grows.

High Energy Consumption Blockchain solutions consume excessive energy due to their intensive computing power requirements. The proof-of-work (PoW) consensus mechanism, commonly used in blockchain networks, requires extensive energy resources to secure and validate the distributed ledger due to its decentralized nature. This can be a significant burden on infrastructure costs and have adverse environmental impacts.

Data Privacy Maintaining data privacy is a significant challenge, particularly in public and permissionless blockchains. The immutability of blockchains can lead to the public storage of sensitive or private information, potentially violating data privacy laws. Additionally, reliance on third-party service providers may expose sensitive data to unauthorized access, becoming potential points of failure or compromise.

We can consider the following potential solutions to address the challenges in implementing blockchain solutions for securing robotic networks:

Heterogeneous Robots and Hybrid Approach By incorporating robots with varying computational capabilities in the network, the computational overhead of blockchain processing can be distributed more efficiently. Delegating blockchain computations to more powerful robots can help mitigate communication delays and ensure that the system operates optimally. Additionally, the implementation of hybrid approaches that combine blockchain for security-relevant data ("on-chain") with traditional local processing for other data ("off-chain") can optimize the overall computation performance.

Scalability Solutions To tackle the scalability issue of blockchain technology, adopting better consensus protocols and implementing sharding techniques (e.g., parallel execution of small sets of blocks) can significantly improve transaction processing speed and throughput. Utilizing nested blockchains or layer 2 solutions can help scale the system efficiently by adding another layer on top of the existing blockchain network. Furthermore, state channels can be employed to handle more transactions per second while reducing the burden on the main blockchain. An alternative approach can also be utilized to reduce blockchain data by transferring them on-demand or storing only hash values for these data [24].

Storage Solution We can address the issue of storage requirements by limiting the amount of storage needed for each robot in a network and its communication needs. For instance, we can use tokens in blockchain as costs to limit the number of transactions issued by robots. This imposes a strong bound on the amount of memory needed by robots to store the chain of blocks, which limits the maximum memory that needs to be installed for them [11]. This technique of using tokens for sending messages by each robot can strike a balance between storage requirements and network security.

Energy Consumption Reduction A significantly less resource-intensive consensus mechanism, namely, proof-of-stake (PoS), compared to POW, can be implemented in [32] in multi-robot settings. Another consensus mechanism such as proof-of-authority (PoA) can be used for reducing energy consumption to implement blockchains. These alternative consensus mechanisms are more suitable for resource-constrained robotic networks while ensuring efficient validation and security. A recent work [6] that combines two principal components of blockchain computations, Merkle tree (MT) and PoW, can optimize energy consumption.

Secure Technologies and Protocols Private or permissioned blockchains, as well as privacy-focused technologies like zero-knowledge proofs (allow users to prove the validity of a statement without revealing the underlying data), can offer solutions for data privacy [40]. Encryption or hashing can be used to protect sensitive data from unauthorized access. Secure multi-party computation (MPC) protocols can allow computations to be performed on encrypted data without revealing the data's contents, thus preserving privacy.

7 Conclusion and Future Directions

In this chapter, we have introduced a consensus algorithm employing a committee of robots within a network to detect spoofed client robots or compromised servers. This algorithm relies on transferred blockchain data to achieve its goal. Our simulation results demonstrate that our approach makes a robot network resilient against the spoofing attack. We believe that we have just scratched the surface of leveraging blockchain for detecting a cyberattack within a robot network. This effort paves the way for several interesting future research directions as detailed below.

In the future efforts of this stream of research, we will evaluate the vulnerabilities of our approach by utilizing machine learning techniques to learn the characteristics of compromised robots under various attack scenarios on a secure network. This will enable us to develop solutions that enhance the attack resilience of our approach for robot networks. Additionally, we intend to conduct real-world tests with programmable drones acting as client robots and a ground vehicle as the server robot to further validate the effectiveness and practicality of our approach. Furthermore, we will investigate a scalable approach for data integrity in multi-robot information gathering from a changing geographic area under communication and resource constraints, while a malicious robot attempts to tamper with data collected from the underlying area [31].

References

1. Alam, T., Taylor, J., Taylor, J., Badsha, S., Shahid, A.R.B., Kayes, A.: Leveraging blockchain for spoof-resilient robot networks. In: Proceedings of Internation Conference on Intelligent Robotics and Applications, pp. 207–216 (2020)
2. Christidis, K., Gupta, V.: Blockchain for iot security: a review. Blockchain Journal **2**(1), 5–20 (2019)
3. Davidson, D., Wu, H., Jellinek, R., Singh, V., Ristenpart, T.: Controlling UAVs with sensor input spoofing attacks. In: Proceedings of the 10th USENIX Workshop on Offensive Technologies (2016)
4. Desai, A., Ostrowski, K., Gazi, V.: Leader–follower formation control of mobile robots: a comparative study. Int. J. Robot. Autom. **27**(1) (2012)
5. Douceur, J.R.: The sybil attack. In: Proceedings of International Workshop on Peer-to-Peer Systems, pp. 251–260 (2002)
6. Escobar, C.C., Roy, S., Kreidl, O.P., Dutta, A., Bölöni, L.: Toward a green blockchain: engineering merkle tree and proof of work for energy optimization. IEEE Trans. Network Serv. Manage. **19**(4), 3847–3857 (2022)
7. Evans, D.: The Internet of Things: how the next evolution of the internet is changing everything. In: Cisco White Paper (2011)
8. Farooq, U., Taylor, H.: Securing iot robotics: challenges and potential solutions. J. Cybersecur. **8**(2), 100–119 (2019)
9. Fernandes, A., Phillips, C.: Sensor vulnerabilities in internet of robotic things: a comprehensive review. IoT Journal **7**(5), 320–335 (2019)
10. Ferrer, E.C.: The blockchain: a new framework for robotic swarm systems. In: Proceedings of the Future Technologies Conference, pp. 1037–1058 (2018)
11. Ferrer, E.C., Jiménez, E., Lopez-Presa, J.L., Martín-Rueda, J.: Following leaders in byzantine multirobot systems by using blockchain technology. IEEE Trans. Robot. **38**(2), 1101–1117 (2021)
12. Fischer, M., Lynch, N., Paterson, M.: Impossibility of distributed consensus with one faulty process. J. ACM (JACM) **32**(2), 374–382 (1985)
13. Gil, S., Baykal, C., Rus, D.: Resilient multi-agent consensus using Wi-Fi signals. IEEE Control Syst. Lett. **3**(1), 126–131 (2018)
14. Gil, S., Kumar, S., Mazumder, M., Katabi, D., Rus, D.: Guaranteeing spoof-resilient multi-robot networks. Auton. Robot. **41**(6), 1383–1400 (2017)
15. Zhang, Q., Yang, L.T., Chen, Z., Li, P.: A survey on deep learning for big data. Inf. Fusion. **42**, 146–157 (2018). https://doi.org/10.1016/j.inffus.2017.10.006, https://www.sciencedirect.com/science/article/pii/S1566253517305328. ISSN 1566-2535.
16. Johnson, P., Hernandez, G.: Consensus mechanisms in multi-agent robotic systems. Adv. Robot. **42**(4), 300–315 (2020)
17. Kumar, S., Iyengar, R.: The confluence of blockchain and IoT: challenges and solutions. Blockchain Tech Review **5**(1), 10–25 (2021)
18. LeBlanc, H.J., Zhang, H., Koutsoukos, X., Sundaram, S.: Resilient asymptotic consensus in robust networks. IEEE J. Sel. Areas Commun. **31**(4), 766–781 (2013)
19. Li, X., Smith, J.: Iot-driven robotics: opportunities and challenges. J. Robot. Syst. **37**(1), 5–20 (2020)
20. Ma, H., Hönig, W., Kumar, T.S., Ayanian, N., Koenig, S.: Lifelong path planning with kinematic constraints for multi-agent pickup and delivery. In: Proceedings of the AAAI Conference on Artificial Intelligence, vol. 33, pp. 7651–7658 (2019)
21. Miller, R., Lee, J.: Collaborative multi-agent robotic systems: consensus, coordination, and communication. Int. J. Robot. Res. **41**(7), 900–920 (2019)
22. Nakamoto, S.: Bitcoin: A peer-to-peer electronic cash system. Tech. rep. (2019)
23. Nguyen, M., Khoa, D.: Blockchain-driven access control in IoT devices. IoT World **9**(3), 200–215 (2020)
24. Nishida, Y., Kaneko, K., Sharma, S., Sakurai, K.: Suppressing chain size of blockchain-based information sharing for swarm robotic systems. In: Proceedings of International Symposium on Computing and Networking Workshops, pp. 524–528 (2018)

25. Patel, M., Singh, D.: Iot-blockchain integration: current landscape and future directions. Int. J. Adv. IoT **7**(4), 420–440 (2021)
26. Python Robotics. https://github.com/AtsushiSakai/PythonRobotics. Accessed on April 5, 2020
27. Queralta, J.P., Westerlund, T.: Blockchain-powered collaboration in heterogeneous swarms of robots. Frontiers in Robotics and AI (2020)
28. Rahman, F., Malik, R.: Blockchain in industrial IoT: securing data exchanges. Industrial IoT Journal **10**(3), 55–75 (2022)
29. Renganathan, V., Summers, T.: Spoof resilient coordination for distributed multi-robot systems. In: Proceedings of the International Symposium on Multi-Robot and Multi-Agent Systems, pp. 135–141 (2017)
30. Reyna, A., Martín, C., Chen, J., Soler, E., Díaz, M.: On blockchain and its integration with IoT challenges and opportunities. Futur. Gener. Comput. Syst. **88**, 173–190 (2018)
31. Samman, T., Dutta, A., Kreidl, O.P., Roy, S., Bölöni, L.: Secure multi-robot information sampling with periodic and opportunistic connectivity. In: Proceedings of the IEEE International Conference on Robotics and Automation (ICRA), pp. 4951–4957 (2022)
32. Samman, T., Spearman, J., Dutta, A., Kreidl, O.P., Roy, S., Bölöni, L.: Secure multi-robot adaptive information sampling. In: Proceedings of IEEE International Symposium on Safety, Security, and Rescue Robotics (SSRR), pp. 125–131 (2021)
33. Sargeant, I., Tomlinson, A.: Modelling malicious entities in a robotic swarm. In: Proceedings of IEEE/AIAA Digital Avionics Systems Conference, pp. 7B1–1–7B1–12 (2013)
34. Saulnier, K., Saldana, D., Prorok, A., Pappas, G.J., Kumar, V.: Resilient flocking for mobile robot teams. IEEE Rob. Autom. Lett. **2**(2), 1039–1046 (2017)
35. Shepard, D.P., Bhatti, J.A., Humphreys, T.E., Fansler, A.A.: Evaluation of smart grid and civilian UAV vulnerability to GPS spoofing attacks. In: Proceedings of Radionavigation Laboratory Conference (2012)
36. Sicari, S., Rizzardi, A., Grieco, L., Coen-Porisini, A.: Security, privacy and trust in internet of things: the road ahead. Comput. Netw. **76**, 146–164 (2015)
37. Strobel, V., Castelló Ferrer, E., Dorigo, M.: Managing byzantine robots via blockchain technology in a swarm robotics collective decision making scenario. In: Proceedings of the International Conference on Autonomous Agents and MultiAgent Systems, pp. 541–549 (2018)
38. Strobel, V., Castelló Ferrer, E., Dorigo, M.: Blockchain technology secures robot swarms: a comparison of consensus protocols and their resilience to byzantine robots. Front. Rob. AI **7**, 54 (2020)
39. Strobel, V., Dorigo, M.: Blockchain technology for robot swarms: a shared knowledge and reputation management system for collective estimation. In: Proceedings of International Conference on Swarm Intelligence, pp. 425–426 (2018)
40. Talukder, S., Carbunar, B.: A study of friend abuse perception in facebook. ACM Trans. Social Comput. **3**(4), 1–34 (2020)
41. Turner, L., Gonzalez, R.: Spoofing attacks in iot-driven robotics: an exploratory study. Journal of Cyber Physical Systems **3**(2), 150–165 (2021)
42. Wheeler, T., Bharathi, E., Gil, S.: Switching topology for resilient consensus using Wi-Fi signals. In: Proceedings of the International Conference on Robotics and Automation, pp. 2018–2024 (2019)
43. Wing—A commercial drone delivery service. https://wing.com. Accessed on June 30, 2020
44. Zhang, H., Wei, L.: Smart homes and blockchain: a practical deployment perspective. J. Smart Syst. **6**(2), 50–65 (2020)
45. Zhao, L., Huang, W.: Tailoring blockchain's consensus for IoT: a review. IoT and Blockchain Journal **3**(2), 35–50 (2021)

Part IV
Enhancing Scalability with Sharding in Distributed Replication Systems

Sharding Distributed Replication Systems to Improve Scalability and Throughput

Siamak Solat and Farid Nait-Abdesselam

Abstract Most existing Byzantine fault-tolerant (BFT) algorithms are very slow and are not designed for large sets of participants trying to reach a consensus. Hence, distributed replication systems that use consensus mechanisms to process clients' requests have major limitations and problems in scalability, throughput, and performance. Such problems are mainly due to the time and message complexity of the consensus algorithms used in such systems. Such limited scalability and low throughput can be significantly improved by using sharding approach as a technique for partitioning a state into multiple shards, each of which is handled by a subset of the network in parallel. Use of sharding for replication systems is inspired by sharding in databases. Sharding has already been implemented in several blockchain-based replication systems, and although it has shown remarkable potential to improve performance and scalability, current sharding techniques have several significant scalability and security issues. In this chapter, we detail the main challenges in the sharding approach and also explain how the sharding approach can enhance the scalability and throughput of distributed replication systems. We also review the most notable sharding protocols designed for distributed replication systems.

1 Introduction

Distributed replication means maintaining a copy of the same data on multiple machines, each of which is called a replica, all of which are connected through

---

S. Solat (✉)
Université Paris Cité, Paris, France

ENGIE, Laboratory for Computer Science and Artificial Intelligence, Paris, France

F. Nait-Abdesselam
Université Paris Cité, Paris, France

University of Missouri Kansas City, Kansas City, MO, USA

N. El Madhoun et al. (eds.), *Building Cybersecurity Applications with Blockchain and Smart Contracts*, Signals and Communication Technology,
https://doi.org/10.1007/978-3-031-50733-5_5

a network [1]. Replication is a basic characteristic of various distributed storage systems and one of the key mechanisms for achieving fault tolerance: if a copy of the data gets faulty, the other copies are still available on other replicas. There might be various motives to replicate data [1]:

- Keeping data geographically close to the users to decrease latency.
- Allowing the network to continue working even if some of the components have failed to improve better availability.
- Scaling out the number of nodes processing queries and requests to enhance throughput.

If the data does not change, replication is simple because it only needs to copy the data once per node; hence, the main challenge in replication is managing data changes, where the data is called dynamic or transactional, that is, the data is frequently modified after being stored in the database [2].

Replication systems that use consensus mechanisms to process clients' requests have major limitations and problems in scalability, throughput, and performance. Such problems are mainly due to the time and message complexity of the consensus algorithms used in such systems. The scalability problem means that system performance and throughput slows down as the number of network nodes increases. Such problems are mainly due to the message complexity of the consensus algorithms. For example, the message complexity of PBFT [3] is $O(n^2)$ and that of Paxos [4] and Raft [5] is $O(n)$. While PBFT is both crash and Byzantine fault-tolerant, Paxos and Raft are only crash fault-tolerant. These message complexities can even be exacerbated with the presence of faulty nodes, when the faulty leader/primary node must be replaced through a view-change process.[1] For example, in the case of PBFT and Paxos, when the leader/primary node fails, the message complexity is exacerbated to $O(n^4)$ and $O(n^2)$, respectively [6]. Figure 1 depicts the number of required message exchanges between nodes in the PBFT consensus algorithm. In any case, as the number of nodes in the network increases, the average processing time of clients' requests also increases, which ultimately leads to a great limitation for scaling up the network. Figure 2 depicts how the mean time to process clients' requests increases as the number of nodes in the network increases when using the Paxos or PBFT consensus mechanisms.

Even by replacing classical consensus mechanisms with proof-of-work (PoW) on networks similar to Bitcoin [8], there are still limits to the scalability, performance, and throughput, as the throughput of the Bitcoin network is only about ≈7–10 transactions per second [9]. Albeit, in general, there is a controversy and a difference of opinion in recognizing PoW as a consensus because there is a belief that it does not have the required properties of a consensus mechanism [10–13].

[1] In a consensus mechanism, a view-change means switching to a new leader node. The view-change as an algorithm for choosing a new leader to collect information and propose it to processor nodes is the epicenter of a replication system [6].

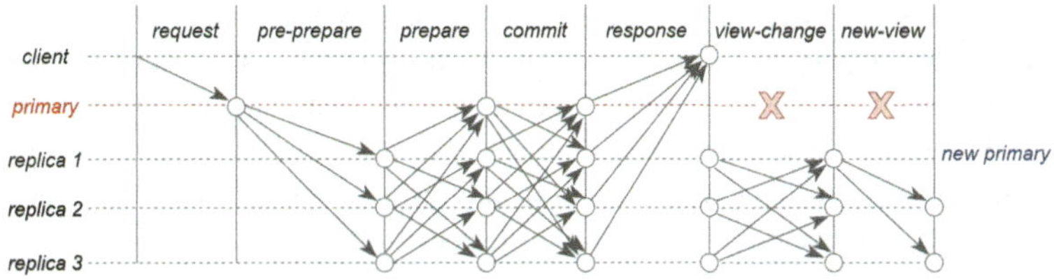

Fig. 1 The PBFT consensus message complexity where a primary node fails and a change-view with additional message exchange is required, so that for f leader failures the message complexity increases to $O(f.n^3)$

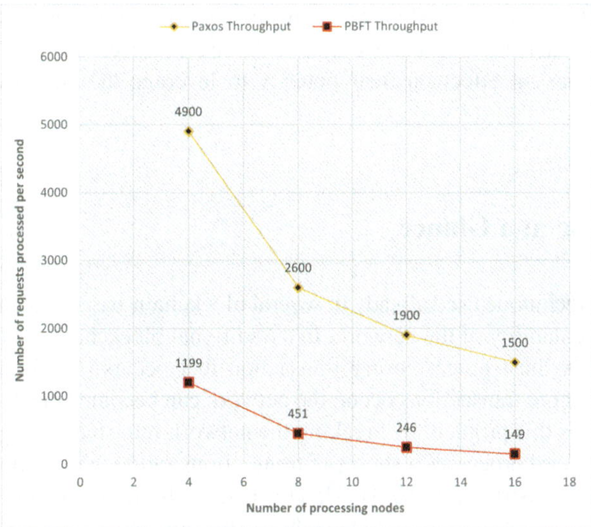

Fig. 2 Throughput of a network that uses Paxos or PBFT consensus decreases drastically, as the number of nodes increases [7]

The problem of scalability becomes very important and crucial when the network is open or so-called permissionless, because no permission from any privileged entity is required to create processor nodes and participate in processing requests or to create client nodes and sending requests (see Definitions 1 and 2). This is why platforms such as Hyperledger [14] use permissioned networks to control the number of nodes by mandating the need for permission from some privileged entity to create processor nodes or send requests in order to limit the size of the network; otherwise, by increasing the number of nodes, the throughput of the network decreases dramatically due to time complexity of the consensus mechanism used for processing clients' requests.

Definition 1 *Permissionless Network:* A replication system is permissionless or public if participation in the submission and the process of transactions is permitted to everyone. There is no special permission for submitting transactions beyond the possession of some way to pay transaction fees. Anyone is also permitted to be a validator[2] to participate in the processing of transactions. This must be in actual practice accessible to everyone who makes a reasonable attempt to earn it. Everyone who sends validly signed transactions to the network should be capable of fairly expecting the network to execute without having to concern that some particular group or entity can decide to prohibit their transactions in particular.

Definition 2 *Permissioned Network:* A replication system is permissioned or private if it does not permit open participation in either submitting or processing transactions, such that sending a transaction needs some permission beyond mere possession of some way to pay transaction fees or participants cannot fairly expect the network to resist censorship, meaning that not all participants have practical guarantee that their transactions would not be discriminated against in a way that considerably has an effect on their potency to leverage the network and get its profits.

2 Sharding at a Glance

Sharding is a technique used already in several blockchain-based systems in order to increase the scalability of the network. In a traditional blockchain system, all nodes on the network must process every transaction that occurs on the network. This means that as more transactions occur, the network can become congested and slow. Sharding solves this problem by breaking the network into smaller segments called shards. Each shard processes a subset of transactions rather than all transactions on the network. By distributing the workload across multiple shards, the network can handle more transactions per second and hence become more scalable. In a sharded replication system, each node is responsible for processing transactions only in its assigned shard. This reduces the computational requirements for each node and makes it easier for new nodes to join the network. Sharding is still an area of active research and development in distributed replication networks, but it has the potential to significantly improve the performance and scalability of blockchain networks. Considering the limitations and obstacles in consensus algorithms for scaling, one of the main reasons for such low throughput in replication systems that use consensus mechanisms is the serial processing approach, where each client request is processed by all processor nodes that leads to a significant decrease in system performance and throughput in a redundancy approach. In contrast to the serial processing, sharding as a parallelism approach is already employed in several replication systems and has shown a notable capability to improve the performance and scalability, yet, the current sharding techniques have several remarkable problems detailed in Sect. 1.

[2] Another term for the nodes in the network which process transactions.

Replication systems such as Bitcoin, Ethereum [15], and Hyperledger that use consensus mechanisms—both classical and so-called "proof-of-x" techniques[3]—to process requests and transactions have low throughput, performance, and scalability. The only reason why non-sharded Ethereum nodes can still store the entire state (or the whole replication) is that Ethereum only processes around 15 transactions per second [17]. Once a system processes thousands of transactions per second, the state will explode, since transactions do leave a trace on the state [18]. A common way of dealing with clients' requests in such systems is serial processing approach, where all the requests are processed by all the processor nodes in the network, and hence, by joining new nodes to the network, the total request processing capacity of the system gradually will decrease. Networks that use classical consensus—either with linear or quadratic message complexity—to process clients' requests lead to increased processing coordination costs [19]. On the other hand, networks that use proof-of-work as an alternative to classical consensus algorithms face the same problem by increasing the computing power of the entire network. Even if the number of processor nodes gets limited by a centralized approach and using a privileged entity in a permissioned network, by increasing the rate of clients' requests, the processor nodes hardware performance is still limited, causing significant latency in response to the clients [19]. One of the proposed solutions to this problem is using sharding technique by dividing the network into multiple smaller groups, each of which handles a part of clients' requests. Several protocols have been already proposed based on sharding technique. We describe briefly some of them in Sect. 3.

The sharding technique can be divided into two general types [20]:

- Processing sharding
- Storage/state sharding

For example, Zilliqa [9] is not a state sharding protocol, as each node holds the entire stored replicated data state to be able to process transactions or clients' requests, while other solutions like Omniledger [21] and RapidChain [22] feature state sharding, where each shard holds a subset of the stored replicated data state. In most of the cases, storage/state sharding typically brings us processing sharding as well. To the best of our knowledge, actually there is no protocol that uses storage sharding without processing sharding.

The rest of this chapter is organized as follows. In Sect. 2.1, we describe the most important challenges in the sharding of distributed replication systems, which are divided into the following four subsections: in Sect. 2.1.1, we explain why most current sharding protocols use a random assignment approach for allocating and distributing nodes between shards due to security reasons. In Sect. 2.1.2 we detail how a transaction is processed in a sharded replication system, based on current sharding protocols. In Sect. 2.1.3 we describe how a shared-ledger

[3] Various methods of using blockchain systems to prove something in a way that is cryptographically verifiable [16].

imposes additional scalability limitations and security issues on the network. And in Sect. 2.1.4, we explain why cross-shard or inter-shard transactions are undesirable and more costly, due to the problems they cause, including atomicity failure and state transition challenges, along with a review of proposed solutions. And, in Sect. 3, we review some of recent works that utilize sharding techniques for replication systems.

2.1 Sharding Challenges

Sharding as a parallelism approach has already been implemented in several distributed replication systems and has shown remarkable potential to improve performance and scalability; nevertheless, current sharding techniques face several challenges. We describe the most important of them below.

2.1.1 Distributing Nodes Between Shards

Most current sharding protocols use a random assignment approach for allocating and distributing nodes between shards due to security reasons. We explain why this approach is employed in most sharding protocols using the following example: assume in a non-sharded replication system, there are in total ten replicas, two of which are Byzantine and also know each other as the members of a cyber-attacker group, that is, they are *colluded* replicas, as depicted in Fig. 3. If the consensus is PBFT, the network can remain safe if the number of nodes, n, is greater than or equal to $3f + 1$, where f is the number of Byzantine or faulty nodes. We then divide the network into two shards, in such a way that the replicas are permitted to choose which shard to assign. Obviously, two Byzantine replicas prefer to be the member of the same shard in order to be able to dominate that shard. Hence, in most sharding protocols, the assignment of replicas between shards is done in a random manner in order to defeat the security problem because in the case of using a random assignment approach, the probability that all the members of an attacker group or colluded replicas are assigned to the same shard is considerably reduced.

2.1.2 How to Process Transactions in Sharded Replication Systems

In this section, we detail how a transaction is processed in a sharded replication system, based on current sharding protocols. In state/storage sharding, processed transactions are stored in separate shards. With state sharding, each node has a shard it is assigned to in such a way that at any given moment of time, the state of the stored replicated data is split between shards in a way known to all nodes.

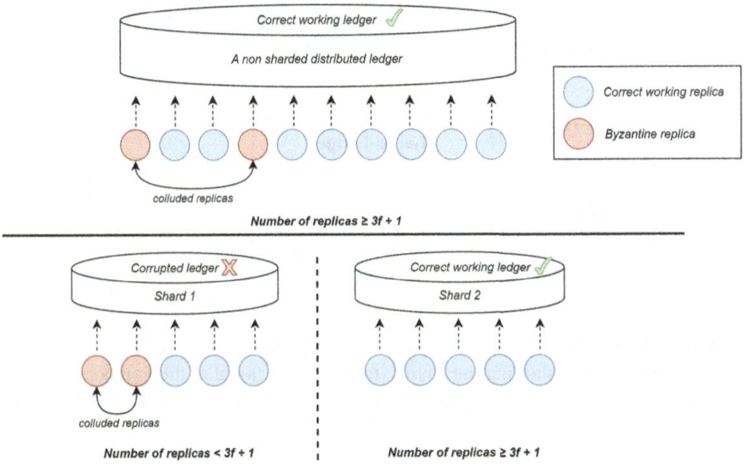

Fig. 3 Most current sharding protocols use a random assignment approach for allocating and distributing nodes between shards due to security reasons

In other words, all the nodes—and everything else that is stored in state—is split between the shards in some way known to all nodes. Each committee is assigned to a particular shard, which is responsible for every particular subset of the state. A transaction is affecting some nodes in the network, meaning that, if a client node $n_{c\alpha}$ makes a transaction and sends some token to another client node $n_{c\beta}$, both nodes are affected by this transaction, so that the token balance[4] of $n_{c\alpha}$ decreases, while that of $n_{c\beta}$ increases. Each node is assigned to exactly one shard, and the transaction is processed by the committee, and only by the committee, that is responsible for the subset of state that the transaction is affecting. If the nodes affected by the transaction are assigned to the same shard, the transaction is an intra-shard transaction, but if each of the nodes are assigned to different shards, a cross-shard or inter-shard transaction occurs such that each participating shard has access to only part of the transaction data for processing and consequently processing an inter-shard transaction is more complex than an intra-shard transaction.

2.1.3 Challenges with Shared-Ledger

Some sharding protocols use a shared-ledger between shards that perform critical coordination mechanisms. Such a shared-ledger imposes scalability limitations and additional security challenges on the system, which we describe in detail.

[4] The number of crypto-tokens that a node holds.

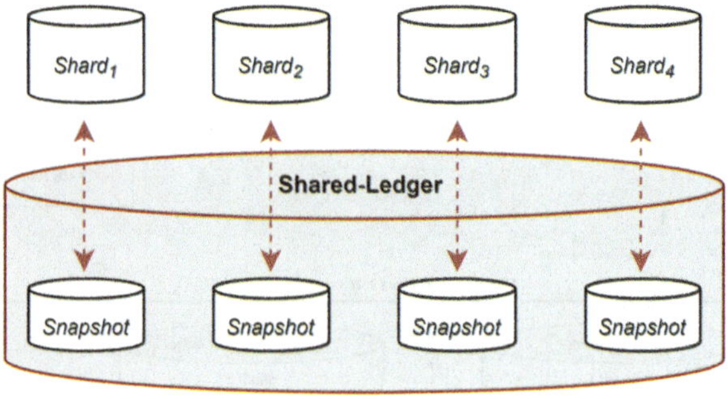

Fig. 4 A shared-ledger doing some bookkeeping computation, such as assigning the processor nodes to the shards, or snapshotting the shards, that is proportional to the number of shards in the system and imposes significant constraints on network scalability, so that the number of shards is naturally limited [19]

2.1.3.1 Scalability Issues Due to Shared-Ledger

The sharding is often advertised as a solution enabling linear scalability, meaning that as the number of nodes in the network increases, the throughput of the system increases at an almost linear rate [23]. While it is in theory possible to design such a sharding protocol, any solution that uses the concept of the shared-ledger between the shards cannot achieve such scalability. Some sharding protocols such as Ethereum 2.0 [24], PolkaDot [25], or Cosmos-Hub [26] use a common ledger to do some bookkeeping computation, such as assigning processing nodes to shards, or snapshotting shard chain blocks, that is proportional to the number of shards in the network [19]. This shared-ledger is called by different names, for example, in sharded Ethereum it is called "Beacon" chain, or "Relay" chain in PolkaDot, or "Cosmos-Hub" in Cosmos protocol, but in this article we call this ledger a "shared-ledger." Since a shared-ledger is itself a ledger with computation bounded by the computational capabilities of nodes operating it, the number of shards is naturally limited [19]. Figure 4 depicts this situation, in which a shared-ledger holds snapshots of data from all shards and ultimately imposes significant constraints on network scalability. Although the participating nodes in shared-ledger keep only the necessary data, however, by growing the network and increasing the number of shards, this ledger encounters the scalability problem, resulting in the scalability limitations of the whole system.

2.1.3.2 Security Issues Due to Shared-Ledger

Besides the scalability problem, if the nodes operating the shared-ledger become Byzantine, then this shard is able to infect the whole system, as crucial tasks such as assigning the nodes between the shards is done by this privileged shard. In other words, a faulty shared-ledger occupied by Byzantine nodes is able to infect a significant part of the system and put it at risk of collapse because a shared-ledger is mission-critical and any bug in it could compromise the whole network [27].

2.1.4 Challenges with Cross-Shard Transactions

In a cross-shard transaction, each participating shard has access to only part of the transaction data for processing, and hence the cross-shard transactions need expensive inter-shard coordination to ensure state consistency and considerably limit the performance of the system. Such circumstances make cross-shard transaction processing more complicated, more complex, and therefore more expensive than an intra-shard transaction. Below we describe some of the challenges a system can face due to cross-shard transactions, but before that, it is important to know two general approaches by which a cross-shard transaction can be processed.

- **Synchronous approach:** In synchronous cross-shard transaction processing, new states (or blocks) that contain state transitions concerning a transaction are all generated simultaneously. In blockchain-based protocols, new states are the same as new blocks. A cross-shard transaction obviously affects a set of shards as explained in Sect. 2.1.2, and synchronous cross-shard transaction must be included at the same block height in all of affected shards. Transactions within a block must be arranged in their hash order to ensure a canonical order of execution. This approach requires a high level of coordination between shards, which leads to higher message complexity and also exacerbates the time it takes to create new states (or blocks), such that blocks are produced simultaneously and all state transitions of a transaction occur at the same block height. With the synchronous approach, blocks are generated as fast as the slowest shard, and hence intra-shard transactions are sacrificed for the communication overhead of cross-shard transactions.
- **Asynchronous approach:** In asynchronous cross-shard transaction processing, each shard generates new states (or blocks) independently, but to ensure atomicity, a shard may need to lock a state transition to ensure that the state is committed on another shard which leads to higher transaction latency, that is, in asynchronous approach, blocks are generated more frequently than synchronous strategy, but on the other hand processing time of a cross-shard transaction might be exacerbated.

To the best of our knowledge, although there are considerable discussions on the synchronous approach, such as [28, 29], there is still no notable sharding protocol that uses a synchronous strategy for cross-shard transactions processing. Hence, we focus more on the challenges of the asynchronous approach.

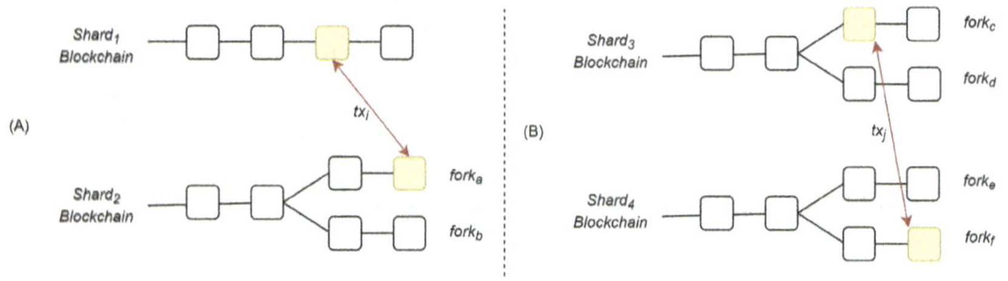

Fig. 5 (**a**): tx_i is a cross-shard transaction between $shard_1$ and $shard_2$, where a fork has occurred. If $fork_a$ in $shard_2$ as a part of transaction tx_i is in canonical/main chain, then tx_i gets finalized; otherwise an atomicity failure has occurred. (**b**): tx_j is a cross-shard transaction between $fork_c$ in $shard_3$ and $fork_f$ in $shard_4$. If both $fork_c$ and $fork_f$ are in canonical/main chain, then tx_j gets finalized. If both forks become abandoned, then tx_j becomes fully abandoned that is no conflict and the situation is fine. But if one of these forks becomes canonical/main chain, while another one is abandoned as a part of forked chain, then an atomicity failure has occurred

2.1.4.1 Atomicity Failure

In this section we detail how an atomicity failure is possible to occur when using a sharded blockchain-based network and while a cross-shard transaction is made. A transaction normally includes two parts: credit (token-receiving) and debit (token-sending); assume that each part is executed in a different shard as an inter-shard transaction. This cross-shard transaction must be either committed in both shards or aborted in both shards; otherwise an atomicity failure occurs. In other words, the result of a transaction must either be a successful commit, so that all changes are made permanent and durable, or an abort, that is, all changes are rolled back, undone, or discarded. This feature is called atomicity. In an asynchronous processing of a cross-shard transaction if in one or both shards a fork occurs, then if the chain of either credit or debit parts becomes aborted as a part of forked chain while the other part still being in canonical/main chain, an atomicity failure has occurred because a part of transaction is validated in a shard, but another part is abandoned in another shard. Figure 5 depicts such a situation.

2.1.4.2 State Transition Challenge

Another challenge with cross-shard transactions is to abuse this type of transaction in order to turn an invalid data transition into a valid one. Consider Fig. 6 on which $shard_a$ is corrupted, where a group of colluding Byzantine processing nodes created invalid block a_2 such that some tokens have been minted unexpectedly on one of their accounts, i.e., ψ. The Byzantine nodes then generate a valid block a_3 on top of invalid block a_2. Hence, while the transactions in block a_3 are applied correctly, those of block a_2 are not made properly. The Byzantine nodes then make a cross-shard transaction toward $shard_b$, whose all blocks are valid and in a correct situation, by transferring those tokens of account ψ to their another account, ξ. From this moment, the improperly created tokens reside on a fully valid ledger in $shard_b$.

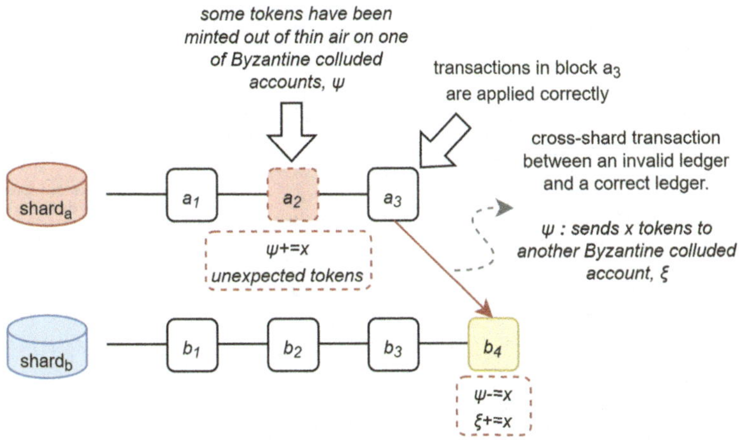

Fig. 6 State transition challenge in sharding

Solutions to State Transition Challenge

First, we introduce the existing solutions for this issue and then analyze their challenges.

- Processing preceding blocks: One of the simplest strategies to defeat the invalid state transition challenge illustrated in Fig. 6 is that processing nodes of shard_b also process the block from which the cross-shard transaction is initiated. This approach would not even work in the example depicted in Fig. 6, because block a_3 appears to be completely valid. As an alternative approach, processing nodes of shard_b should also process some large number of blocks preceding the block from which the cross-shard transaction is initiated. Even this alternative cannot be efficient, as for any number of blocks that are validated by shard_b, the Byzantine nodes in shard_a can generate one more valid block on top of the invalid block that they created.
- Graph-based solution: Another approach to solve the state transition issue in cross-shard transactions is to arrange the shards in an undirected graph so that each shard is connected to several other shards and only cross-transactions between neighboring shards are permitted. This idea is used in [30, 31]. Cross-shard transactions between non-neighboring shards are routed through multiple shards. In this approach, a processing node in each shard is expected to process both all transactions in its own shard and all transactions in all neighboring shards. Figure 7 depicts such a strategy so that shard_b processes not only its own transactions but also the transactions of all its neighbors, including shard_a. Hence, the group of colluding Byzantine processing nodes are not able finalize the cross-transaction performed in Fig. 6, because shard_b processes the entire history of shard_a as well—as its neighbor—leading to the identification of invalid block a_2.

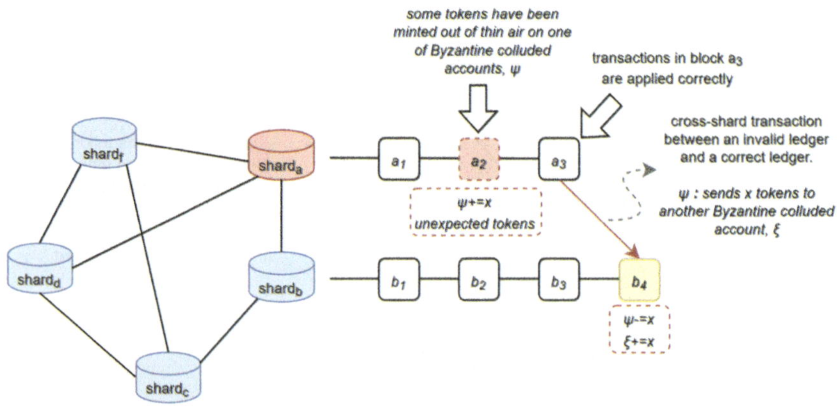

Fig. 7 A graph-based solution can resolve the state transaction problem in some cases

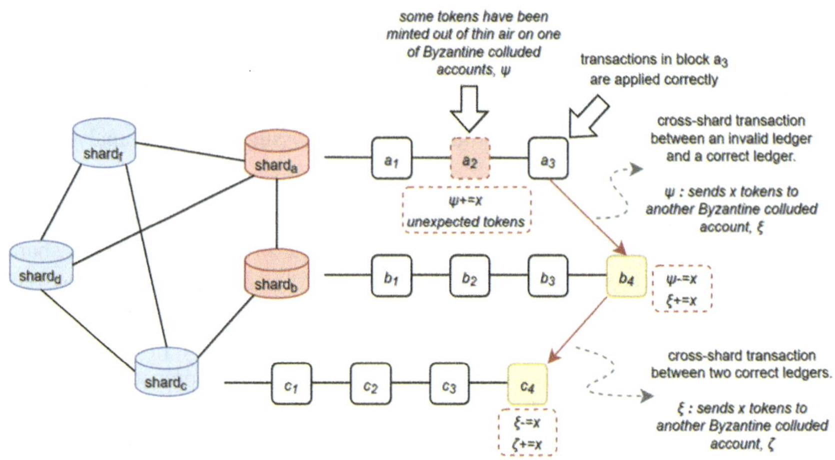

Fig. 8 The graph-based solution is not always able to resolve the state transaction problem

With the graph-based approach, while corrupting one shard no longer leads to an effective attack, corrupting multiple shards can still be problematic. In Fig. 8, a colluding Byzantine processing node that has managed to corrupt both shard_a and shard_b can successfully execute a cross-shard transaction to shard_c with funds originating from invalid block a_2. Shard_c processes the entire ledger of shard_b, but not that of shard_a, and thus it is not able to detect invalid block a_2.

In the following, we describe two more effective solutions to the data transition challenge in cross-shard transactions along with their weaknesses:

- Fisherman: The idea behind the Fisherman approach is that whenever a cross-transaction occurs, if an honest processing node is present in the shard where the invalid block has been created, within a certain period of time called the challenge period, it can prove that the block is invalid. In fact, a Fisherman node is a member of a shard controlled by a group of colluding Byzantine nodes, and if an invalid block is created by the Byzantine group, it is reported by the Fisherman to the token receiving shard with which a cross-shard transaction is made. As an advantage, this approach works as long as there is at least one honest processing node in the part where the invalid block exists. In order to optimize the communication overhead for the receiving nodes, various constructions are used that enable succinct proof of invalidity of malicious blocks.

Although this idea is the dominant approach among today's proposed protocols, it has two following weaknesses: the challenge period must be long enough for the honest processor to prepare the challenge and then thoroughly check whether the block is invalid or not. Considering such a period significantly can reduce the speed of cross-shard transactions.

On the other hand, this approach and challenge periods to process invalid blocks create a potential of a new kind of attacks in the form that colluding Byzantine nodes spam with invalid challenges. A solution for this type of attack is to block some tokens from challenge senders as a deposit or collateral and only return them if the challenge is correct. However, this solution may not be efficient enough, as it may still be profitable for attackers to spam the system and burn its deposits with invalid challenges, for example, when tokens that have been minted out of thin air in invalid block a_2 are more than burned collateral tokens or, in another scenario, in the case of a griffing attack, an attack that does not necessarily benefit the attacker, but the main motive is to make it harder for the victim to use the system, i.e., to cause him grief, as the name implies.

3 Overview of Sharded Distributed Replication Protocols

In this section, we review some of the most considerable recent works that utilize sharding techniques for replication systems. We firstly select two notable sharding protocols to describe more deeply as two general kinds of ecosystems in sharding techniques: Ethereum 2.0 [32] as one of the most used blockchain-based platforms that is a homogeneous multi-chain sharding system and Polkadot [25], as a heterogeneous multi-chain sharding protocol. Unlike two homogeneous blockchains, two heterogeneous blockchains do not have identical architecture and the same characteristics. We then review some other considerable sharding protocols in Sect. 3.3 that each of them tried to improve a part of sharding challenges.

3.1 Sharded Ethereum: A Homogeneous Multi-Chain Protocol

Sharded Ethereum (also known as Ethereum 2.0) [32] is an upgraded version of Ethereum that at the time of writing this article is still in progress and aims at transforming Ethereum into a sharded network. Although there are efficient

Ethereum clients like Parity [33] that can process around 3000 transactions per second, provided the hardware is efficient enough, nevertheless, non-sharded Ethereum network is not enabled to process more than about ≈30 transactions per second [9]. Hence, Ethereum Foundation decided to upgrade the protocol to a sharding-based system including a multi-phase upgrade to improve Ethereum scalability and capacity. In the first phase of the implementation, they set up 64 shards to test the Beacon chain's finality [34]. In the following, we describe the main components as well as major challenges of Ethereum 2.0.

3.1.1 Beacon Chain

The Beacon chain is the main component of the Ethereum 2.0 network. Shard Chains, on the other hand, are ledgers where transactions are executed. Each shard chain has an independent state, so it is only responsible for processing transactions related to that state. The Beacon chain performs important tasks such as tracking information about validators, their stakes, attestations and votes, and also slashing validators if they are found to be dishonest, meaning that dishonest validators lose part of their stakes. The Beacon chain, as well as its underlying protocol, is responsible for administering consensus between validators on the state of the system. Due to the coordination role and the amount of assets managed by the Beacon chain, this ledger is a mission-critical component of the Ethereum 2.0 ecosystem [27], and as explained in Sect. 2.1.3, any bug in this shared-ledger could compromise the whole network (Fig. 9).

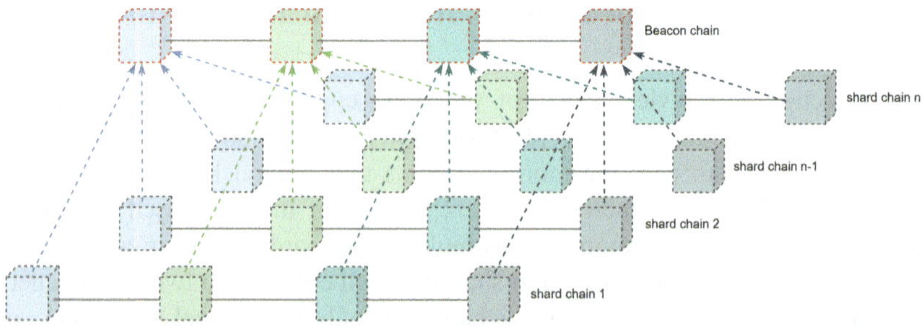

Fig. 9 The figure depicts shard chains, where transactions are executed as well as the Beacon chain, which performs important tasks such as tracking information about validators, their stakes, attestations and votes, and also slashing validators if they are found to be dishonest, meaning that dishonest validators lose part of their stakes. The Beacon chain, as well as its underlying protocol, is responsible for administering consensus between validators on the state of the system

3.1.2 PoS and Block Generation:

In Ethereum 2.0, proof-of-work will eventually be replaced by proof-of-stake (PoS),[5] where blocks are proposed on a fixed schedule [32]. Ethereum 2.0 finalizes batches of blocks based on time periods called epochs [36]. An epoch is defined as some constant number of slots. In Ethereum 2.0, time is measured in slots, and it is defined as some constant number of seconds. In a tentative version, an epoch is considered 64 slots and each slot 12 seconds [37]. It is planned to finalize 32 blocks in each batch during one epoch. With an estimated block time of 12 seconds, finality is expected to take between 6 and 12 minutes [36]. To generate blocks, Ethereum 2.0 uses the random decentralised autonomous organization (RandDAO), which is a slot-based protocol that randomly selects validators for a slot and enforces a fork choice rule for unfinalized blocks [38]. Each validator instance requires 32 ETH (Ethereum cryptocurrency) as stake. Sets of validators are randomly selected and form groups called committees that validate shards on the network. A large number of validators are needed to ensure the validity [39]. A minimum of 111 validators per shard is required to run the network, and 256 validators per shard are required to finalize all shards within one epoch. Hence, with 64 shards planned for the first phase, 16,384 validators are needed [34].

3.1.3 Roles and Terminology in Ethereum 2.0

In Ethereum 2.0, validators participate in the consensus mechanism and are called virtual miners. A proposer is selected pseudo-randomly to propose and build a block for each slot. On the other hand, attesters vote on the proposed blocks of both the Beacon chain and the shard chains. The vote of the validators is called attestation. In most cases, validators are attesters who vote on blocks, and their attestations are recorded on the Beacon chain. Proposers receive a reward if their proposed block gets confirmation of a quorum of attesters. Each attestation has a weight, which is actually the amount of the stake of the validator who writes the attestation [37]. This approach is used for the fork choice rule mechanism that we describe in Sect. 3.1.4. Validators monitor each other and are rewarded for reporting validators who generate conflicting attestations or propose multiple blocks. Each block is proposed by a random block proposer to be added to the Beacon chain in each slot. If, for a given slot, a validator does not see a generated block, or does not receive the block in time, or if the block was generated on a chain that the validator does not recognize as the current chain, it should generate an attestation that the slot is empty by attesting to the block it believes is the head of the chain. That is, a validator should attest to exactly one block per slot, either attesting to the

[5] An alternative to PoW to reduce the computational cost, which was first used in Peercoin protocol [35].

actual block generated by a proposer or making an attestation showing that the slot is empty.

3.1.4 Consensus in Ethereum 2.0

Gasper, which is a combination of the Casper-FFG (Casper the Friendly Finality Gadget) and LMD-GHOST (Latest Message-Driven-Greedy Heaviest Observed Sub-Tree), is designed for a full proof-of-stake-based blockchain system, where a validator's voting power is proportional to their stake (or cryptocurrency) in the system such that instead of using computational power to propose blocks, proposing blocks is essentially free [37]. Unlike Gasper, Casper-FFG is a hybrid PoW/PoS system. It is also based on the consensus theory of Byzantine fault tolerance [40]. In fact, Casper-FFG implements a PoS mechanism as an overlay on top of a PoW ledger to achieve more energy-efficient finality by creating a hybrid consensus model. Casper-FFG is designed to be compatible with a wide range of blockchain protocols with tree-like structure and is a "finality gadget," meaning that is not a fully specified protocol and is designed to be a "gadget" that works on top of a provided blockchain protocol, agnostic to whether the provided chain is proof-of-work or proof-of-stake [37]. Casper-FFG is an algorithm that marks certain blocks in a blockchain as finalized so that participants with partial information can still be fully confident that the blocks are part of the canonical chain of blocks [37]. Both Gasper and Casper-FFG define the concepts of "justification" and "finalization" which are analogous to phase-based concepts in the PBFT literature such as "prepare" and "commit' [37], although in Gasper the "pairs"[6] are justified and finalized rather than the "checkpoint blocks" in Casper-FFG. In order for a block to be "justified," two-thirds of all staked ETH must have voted in favor of including that block in the canonical chain. If another block is "justified" above a "justified" block, that block is upgraded to "finalized" [41]. Also, in Ethereum 2.0, validators are considered as the same replicas in PBFT [37].

3.1.4.1 Fork Choice Rule in Ethereum 2.0

As transactions throughput accelerate, the probability of blockchain being forked also increases. This may include short-term forks and the possibility of various kinds of censorship. Fork choice rule in Ethereum 2.0 is called LMD-GHOST, which stands for Latest Message Driven Greediest Heaviest Observed Subtree. Bitcoin's proof-of-work which uses the longest chain rule as a fork-choice rule that returns the leaf block which is farthest from the genesis block as the "heaviest chain" that has heaviest amount of work and that's why it is also called "most difficult chain," but in Ethereum's proof-of-stake, each attestation has a weight, which is the stake of the

[6] A pair consists of a block and an epoch.

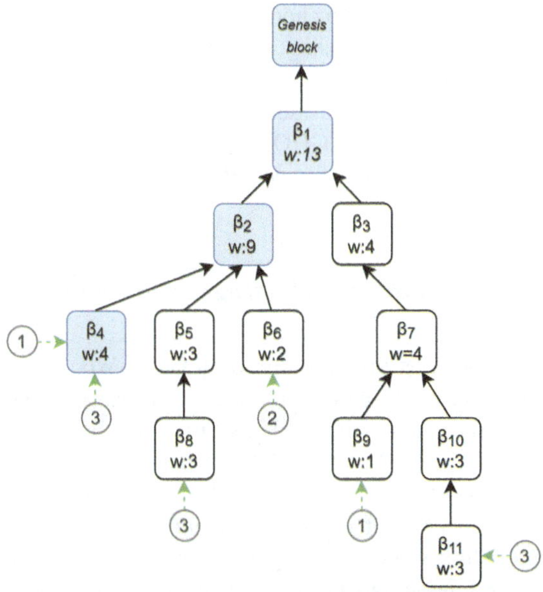

Fig. 10 An example of how to select a subtree using the LMD-GHOST fork choice rule in a view by a validator

validator writing the attestation as a vote by which the fork with the heaviest weight is assumed to be the right one to be the head of the canonical chain. GHOST is a greedy algorithm that grows the blockchain in sub-branches with the "most activity" [37]. LMD-GHOST is a fork-choice rule where validators (participants) attest to blocks to signal support for those blocks, like voting [37]. In LMD-GHOST, we will always end up at a leaf block, which defines a canonical chain [37]. To define LMD-GHOST, it is necessary to first define a concept called weight. Buterin et al. [37] define weight as follows. If assume S to be the set of latest attestations, such that one per validator, the weight of block b is defined as the sum of the stake of validators whose last attestation is either to β or β's descendants. The idea of LMD-GHOST is that at any fork, the subtree of a fork with the heaviest weight is assumed to be the right one, so that it always ends up at a leaf block that defines a canonical chain. Figure 10 illustrates an example of the fork choice rule based on LMD-GHOST.

Each fork choice in Ethereum 2.0 by a validator v is made in a view at a given time t, denoted by $view(v, t)$, as the set of all accepted messages that v has seen so far. In a nutshell, based on the LMD-GHOST fork choice rule, anywhere there is a fork, the heaviest subtree is chosen. In this way, in Fig. 10, the subtree starting with block β_2 is selected because its weight is 9 and greater than the weight of block β_3, which has a weight of 3. Then, similarly, among the three children of block β_2, the block with the highest weight, i.e., block β_4 is selected. Thus, using the view illustrated in Fig. 10, a validator will recognize the blue chain as a canonical chain. This figure is modeled after the original figure in the article "Combining GHOST and Casper" [37] where the LMD-GHOST protocol has been detailed. The main idea of GHOST is to choose a side with more overall support for validators in each fork instead of choosing a subtree that is longer, so that in addition to the number of attestations, the weight of each attestation, which is based on the stake of its validator, is also considered. This idea is heavily inspired by Sompolinsky et al. [42] in the original GHOST paper, but LMD-GHOST adapts the design from its original PoW context to new PoS context [43].

3.2 Polkadot: A Heterogeneous Multi-Chain Protocol

Polkadot was first introduced by Gavin Wood in 2016 as a heterogeneous multi-chain protocol aiming to provide a scalable and interoperable framework for multiple chains with pooled security that is achieved by the collection of components [25]. Polkadot has its own native crypto-token called DOT. Polkadot uses a central chain called the *Relay* chain that communicates with several heterogeneous and independent chains called *parachains*. The Relay chain is responsible for providing shared security to all parachains as well as enabling trustless cross-chain transactions between parachains. The issues Polkadot intends to address are interoperability, scalability, and weaker security resulting from splitting the security power [44]. The Polkadot Relay chain consists of nodes and roles. Nodes are network-level entities that physically run Polkadot software, and roles are protocol-level entities that administer specific purposes. At the network level, Relay chain nodes can participate as either light clients or full nodes. Unlike light clients, which retrieve certain user-relevant data from the network and are not required to be always available because they do not perform a service for others, full nodes retrieve all types of data and store them for a long time and disseminate to others and therefore should be highly available [44]. In addition to data distribution, Relay chain nodes perform specific roles at the protocol level as follows:

- Validators: as the Relay chain full nodes, do the bulk of the security work.
- Nominators: are shareholders who elect the validator candidates. This can be done through a light client without need of any awareness of parachains.

On the other side, parachains can determine their internal network structure but are expected to interact with Polkadot through the following roles:

- Collectors: collect parachain data and send it to the Relay chain. Collectors are selected as defined by parachain and must be its full nodes. Validators interact with parachain collators, but do not need to participate in parachain as a full node.
- Fishermen: as we explained in Sect. 2.1.4.2 how a Fisherman can help for data validity challenges, they, as a full node of parachain, administer additional security checks on the correct functioning of the parachain on behalf of the Relay chain that provides a reward to incentivize the Fishermen.

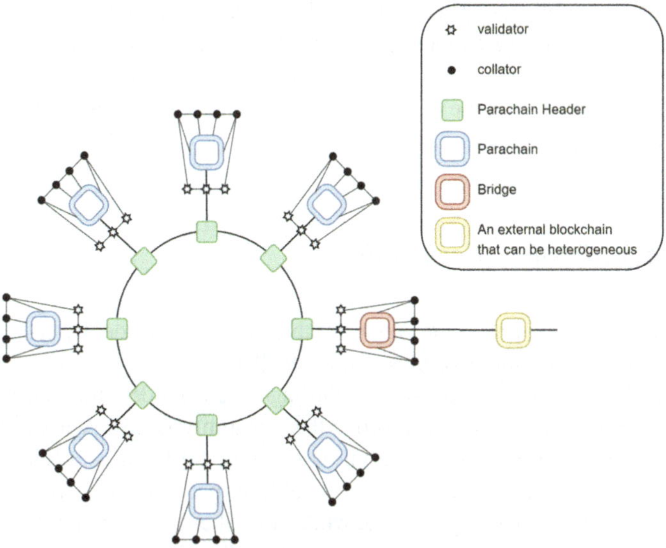

Fig. 11 A high-level view of the Polkadot architecture

In addition to the components listed above, the bridges are intended for compatibility and interaction of the polkadot ecosystem with other external blockchain systems such as Bitcoin, Ethereum, or Tezos [34].

After describing the nodes and roles, we explain more details of how the Polkadot Relay chain protocol works as follows. Collaborators watch the progress of the block-producing and consensus protocols and sign the data building on top of the latest chain block and send it to the validators assigned to their parachain in order to include it into the Relay chain. The parachain validators decide which of the parachain block to support in order to present its relevant data as a parachain next candidate for being added to the next Relay chain block. A block-producing validator makes a set of candidates from all parachains and puts it into a Relay chain block. Validators send their votes on a block and finalize it. All votes are included in the Relay chain blocks. Figure 11 depicts a high-level view of the Polkadot architecture in an example that includes 8 parachains, 24 validators, and 4 collators per parachain, along with a bridge connecting the entire system to another external blockchain network so that two blockchain networks can be heterogeneous. This figure is modeled after the figure in [44], where the Polkadot protocol is thoroughly reviewed.

As a consensus mechanism, Polkadot uses nominated PoS (NPoS) that is a modified version of proof-of-stake. Since NPoS has a deterministic finality, a set of registered validators of bounded size is required. DOT holders can participate in the NPoS consensus as nominators. To register as a nominator, at least 100 DOT is required. The validators' candidacies are visible to all nominators, and each nominator publishes a list of up to 16 candidates it supports, and then the network automatically distributes the stake among the validators. And finally, a certain number of validators that have the most DOT (as stake) are selected and activated. In NPoS, the stake of nominators and validators may be slashed, as a security measure [34].

3.3 Other Sharded Blockchains

In this section, we review some other notable sharding protocols, each of which attempts to ameliorate some of the sharding challenges.

Zilliqa [9] was proposed as a sharding-based model for permissionless blockchain networks in order to improve the scalability issues. A major weakness of Zilliqa is that it shards processing but not storage [19], that is, each node holds the entire stored replicated data state to be able to process transactions [19]. It is worth noting that not sharding by state, while simplifies the system design, imposes a huge limit on the scalability of the system [18]. The only reason why non-sharded Ethereum nodes can still store the entire state (or the whole replication) is that Ethereum only processes around 15 transactions per second [17]. Once a system processes thousands of transactions per second, the state will explode, since transactions do leave a trace on the state [18]. In fact, only supporting processing sharding prevents machines with limited resources from participating in the network, thus curtailing decentralization [20]. Zilliqa uses PoW as an identity registration process as a Sybil attack [45] mitigation mechanism. Zilliqa's consensus core relies on PBFT by improving its efficiency using EC-Schnorr multi-signature as developed in [46, 47].

Elastico [48] is proposed as a sharding-based protocol for permissionless blockchain networks and uniformly partitions network into smaller committees, each of which processes a disjoint set of transactions. According to the results of their experiments to measure the scalability of the network based on PBFT consensus, when network size increases from 40 to 80 nodes (2 times), the latency to reach consensus for each transaction is 6 times longer (e.g., from 3 seconds to 18 seconds), and their experiment when the network has 320 nodes did not terminate after running for 1 hour.

Omniledger [21] is another sharding-based solution, whose construction is close to Elastico, but it brings up some challenges in Elastico and then targets to solve them. In Omniledger, the validators are selected by use of proof-of-work. In this way, they use a sliding window of latest block miners as the validator set. They also utilize proof-of-stake as an alternative Sybil attack-resistant approach for choosing a set of validators, aiming to achieve a more power-efficient consensus. They implemented a prototype in Go language on commodity servers (12-core VMs on Deterlab), and their experimental results show that OmniLedger throughput is 6000 transactions per second with a 10 second consensus latency for 1800 nodes.

RapidChain [22] is another sharding-based blockchain system that can be resilient to faulty nodes up to 1/3 of the total participants and is an epoch-based and two-level-BFT[7] protocol whose network needs a reference BFT committee to perform a distributed randomness generation scheme similar to Elastico and create a public random string for initializing the committees. Their empirical evaluation shows that RapidChain throughput is about 7300 transactions per seconds with network sizes of 4000 nodes with 250 committees.

SharPer [49] is a permissioned blockchain system in which nodes are clustered and each data shard is replicated on the nodes of a cluster so that each cluster maintains only a view of the ledger. In SharPer, the blockchain ledger is formed as a directed acyclic graph (DAG).

Ren et al. [50] introduce a permissioned blockchain and call it "spontaneous sharding" so that the network consists of three parts: (a) individual chains generated by each node to record their own transactions in a first-in-first-out fashion, (b) a main chain for a global shared state that uses PBFT as its consensus algorithm and the blocks consist of abstracts signed by the corresponding nodes. They assume that the abstracts of all genesis blocks are on the main chain. Honest nodes will send abstracts of their newest blocks to the main chain when they observe that their previous abstracts are on-chain, and (c) a validation scheme for validation of the transactions. The transactions on individual chains are arbitrary in the sense that they are neither tamper-proof nor signed. The transactions will be tamper-proof and signed if an abstract of a block that comes after it is contained in the main chain, which they call "confirmed" transactions.

4 Conclusion

In this chapter, we described the most important challenges in the sharding of distributed replication systems, an approach that has already been implemented in several blockchain-based replication systems, and although it has shown remarkable potential to improve performance and scalability, current sharding techniques have several significant scalability and security issues. We explained why most current sharding protocols use a random assignment approach for allocating and distributing nodes between shards due to security reasons. We also detailed how a transaction is processed in a sharded replication system, based on current sharding protocols. We described how a shared-ledger imposes additional scalability limitations and security issues on the network and explained why cross-shard or inter-shard transactions are undesirable and more costly, due to the problems they cause, including atomicity failure and state transition challenges, along with a review of proposed solutions. Finally, we reviewed some of the most considerable recent works that utilize sharding techniques for replication systems. We firstly selected

[7] Byzantine fault tolerance.

two notable sharding protocols to describe more deeply as two general kinds of ecosystems in sharding techniques: Ethereum 2.0 as one of the most used blockchain-based platforms that is a homogeneous multi-chain sharding system and Polkadot, as a heterogeneous multi-chain sharding protocol. We then reviewed some other considerable sharding protocols that each of them tried to improve a part of sharding challenges.

References

1. Kleppmann, M.: Designing data-intensive applications: the big ideas behind reliable, scalable, and maintainable systems. O'Reilly Media, Inc., New York (2018)
2. Stephens, R., Ronald P.: Database Design. Sams Publishing, Carmel (2000)
3. Castro, M., Barbara L.: Practical Byzantine fault tolerance. In: OsDI, vol. 99, No. 1999 (1999)
4. Lamport, L.: The part-time parliament. In: Concurrency: The Works of Leslie Lamport, pp. 277–317 (2019)
5. Ongaro, D., John O.: In search of an understandable consensus algorithm. In: 2014 USENIX Annual Technical Conference (USENIX ATC 14) (2014)
6. Yin, M., et al.: Hotstuff: BFT consensus with linearity and responsiveness. In: Proceedings of the 2019 ACM Symposium on Principles of Distributed Computing (2019)
7. Alqahtani, S., Murat D.: Bottlenecks in blockchain consensus protocols. In: 2021 IEEE International Conference on Omni-Layer Intelligent Systems (COINS). IEEE, New York (2021)
8. Nakamoto, S.: Bitcoin: A peer-to-peer electronic cash system. In: Manubot (2019)
9. Team, Zilliqa: The zilliqa technical whitepaper. Retrieved September 16 (2017) (2019)
10. Ethereum community: Consensus mechanism, Sybil resistance & chain selection. In: PoW and PoS alone are not consensus protocols, but they are often referred to as such for simplicity. https://ethereum.org/en/developers/docs/consensus-mechanisms/#sybil-chain
11. Gü Sirer, Emin: there is a terribly wrong framework emerging around consensus protocols. People think that PoW and PoS are consensus protocols, and that they are the only two consensus protocols out there. This is false (2018). Tweet
12. Zhelezov, Dmitrii. In: PoW, PoS and DAGs are NOT consensus protocols. (2018). https://medium.com/coinmonks/a-primer-on-blockchain-design-89605b287a5a
13. Beyer, S.: Proof-of-Work Is Not a Consensus Protocol: Understanding the Basics of Blockchain Consensus. Medium (2019). https://medium.com/cryptronics/proof-of-work-is-not-a-consensus-protocol-understanding-the-basics-of-blockchain-consensus-30aac7e845c8 (accessed April 1, 2019)
14. Dhillon, V., Metcalf, D., Hooper, M.: The hyperledger project. In: Blockchain Enabled Applications. Apress, Berkeley, CA, pp. 139–149 (2017)
15. Buterin, V.: Ethereum white paper. In: GitHub Repository, vol. 1, pp. 22–23 (2013)
16. The Bitcoin Wiki, Category: Proof-of-x. https://en.bitcoin.it/wiki/Category:Proof-of-x
17. Bez, M., Giacomo F., Tullio V.: The scalability challenge of ethereum: an initial quantitative analysis. In: 2019 IEEE International Conference on Service-Oriented System Engineering (SOSE). IEEE, New York (2019)
18. Limitations of Zilliqa's sharding approach. https://medium.com/nearprotocol/limitations-of-zilliqas-sharding-approach-8f9efae0ce3b
19. Skidanov, A., Illia P.: Nightshade: Near protocol sharding design, 39 (2019). https://nearprotocol.com/downloads/Nightshade.pdf
20. Team, H.: Harmony: Technical Whitepaper (2018)
21. Kokoris-Kogias, E., et al.: Omniledger: a secure, scale-out, decentralized ledger via sharding. In: 2018 IEEE Symposium on Security and Privacy (SP). IEEE, New York (2018)

22. Zamani, M., Mahnush M., Mariana R.: Rapidchain: scaling blockchain via full sharding. In: Proceedings of the 2018 ACM SIGSAC Conference on Computer and Communications Security (2018)
23. Mosakheil, J.H.: Security threats classification in blockchains (2018)
24. Ethereum Sharding. Beacon Chain. https://ethereum.org/en/roadmap/beacon-chain/
25. Wood, G.: Polkadot: vision for a heterogeneous multi-chain framework. White paper 21.2327, p. 4662 (2016)
26. Kwon, J., Ethan B.: Cosmos whitepaper. A Netw. Distrib. Ledgers (2019)
27. Cassez, F., Joanne F., Aditya A.: Formal verification of the ethereum 2.0 beacon chain. In: Tools and Algorithms for the Construction and Analysis of Systems: 28th International Conference, TACAS 2022, Held as Part of the European Joint Conferences on Theory and Practice of Software, ETAPS 2022, Munich, Germany, April 2–7, 2022, Proceedings, Part I. Springer International Publishing, Cham (2022)
28. Vitalek B.: Merge blocks and synchronous cross-shard state execution. https://ethresear.ch/t/merge-blocks-and-synchronous-cross-shard-state-execution/1240
29. Casey D.: Synchronous cross-shard transactions with consolidated concurrency control and consensus (or how I rediscovered Chain Fibers). https://ethresear.ch/t/synchronous-cross-shard-transactions-with-consolidated-concurrency-control-and-consensus-or-how-i-rediscovered-chain-fibers/2318
30. Martino, W., Monica Q., Stuart P.: Chainweb: a proof-of-work parallel-chain architecture for massive throughput. Chainweb whitepaper 19 (2018)
31. Zamfir, V.: Ethereum Sharding Proof of Concept. https://github.com/smarx/ethshardingpoc/tree/a0ec249f3fec61279fcde30b403cefebfb23580d#ethereum-sharding-proof-of-concept
32. Ethereum Shard Chains. https://ethereum.org/en/upgrades/shard-chains/#what-is-sharding
33. Parity Ethereum client. https://github.com/openethereum/parity-ethereum
34. Polkadot Wiki. https://wiki.polkadot.network/docs/getting-started
35. King, S., Scott N.: Peercoin: Peer-to-peer crypto-currency with proof-of-stake. self-published paper, August 19.1 (2012)
36. Ethereum 2.0 Block Time. https://github.com/ethereum/consensus-specs/blob/676e216/specs/phase0/beacon-chain.md#time-parameters
37. Buterin, V., et al.: Combining GHOST and casper. arXiv preprint arXiv:2003.03052 (2020)
38. O'Brien, D., et al.: Final Report of the Exploratory Research Project, Blockchain for Transport (BC4T) (2022)
39. Tennakoon, D., Vincent G.: Dynamic blockchain sharding. In: 5th International Symposium on Foundations and Applications of Blockchain 2022 (FAB 2022). Schloss Dagstuhl-Leibniz-Zentrum für Informatik (2022)
40. Buterin, V., Virgil G.: Casper the friendly finality gadget. arXiv preprint arXiv:1710.09437 (2017)
41. Ethereum Developers Docs Consensus Mechanism Gasper. https://ethereum.org/en/developers/docs/consensus-mechanisms/pos/gasper/
42. Sompolinsky, Y., Aviv Z.: Secure high-rate transaction processing in bitcoin. In: Financial Cryptography and Data Security: 19th International Conference, FC 2015, San Juan, Puerto Rico, January 26–30, 2015, Revised Selected Papers 19. Springer, Berlin (2015)
43. Ethereum 2.0 Phase 0—Beacon Chain Fork Choice. https://github.com/ethereum/annotated-spec/blob/master/phase0/fork-choice.md
44. Burdges, J., et al.: Overview of polkadot and its design considerations. arXiv preprint arXiv:2005.13456 (2020)
45. Douceur, J.R.: The sybil attack. In: International Workshop on Peer-to-Peer Systems. Springer, Berlin (2002)
46. Syta, E., et al.: "Keeping authorities" honest or bust "with decentralized witness cosigning." In: 2016 IEEE Symposium on Security and Privacy (SP). IEEE, New York (2016)
47. Kogias, E.K., et al.: Enhancing bitcoin security and performance with strong consistency via collective signing. In: 25th Usenix Security Symposium (Usenix Security 16) (2016)

48. Luu, L., et al.: A secure sharding protocol for open blockchains. In: Proceedings of the 2016 ACM SIGSAC Conference on Computer and Communications Security (2016)
49. Amiri, M.J., Divyakant A., Amr, E.A.: Sharper: sharding permissioned blockchains over network clusters. In: Proceedings of the 2021 International Conference on Management of Data (2021)
50. Ren, Z., et al.: A scale-out blockchain for value transfer with spontaneous sharding. In: 2018 Crypto Valley Conference on Blockchain Technology (CVCBT). IEEE, New York (2018)

Part V
Data Protection in Distributed Ledger and Blockchain Technologies

Data Protection Challenges in Distributed Ledger and Blockchain Technologies: A Combined Legal and Technical Analysis

Danaja Fabčič Povše, Alfredo Favenza, Davide Frey, Zoltán Ádám Mann, Angel Palomares, Lorenzo Piatti, and Jessica Schroers

Abstract Blockchain and the blockchain-based cryptocurrency Bitcoin revolutionized our ideas of decentralized data and transaction management. Based on and improving on the ideas of blockchain, a variety of distributed ledger technologies (DLTs) have been proposed in recent years. DLTs promise fully decentralized data and transaction management, with wide-ranging applications that go well beyond cryptocurrencies. However, DLTs are also associated with challenges, particularly with respect to compliance with applicable data protection regulations.

This chapter presents a comprehensive analysis of the challenges associated with DLTs' compliance with the General Data Protection Regulation (GDPR) of the European Union (EU). It analyzes the impact of these challenges on different

D. F. Povše
Vrije Universiteit Brussels, Belgium
e-mail: Danaja.Fabcic.Povse@vub.be

A. Favenza
LINKS Foundation, Turin, Italy
e-mail: alfredo.favenza@linksfoundation.com

D. Frey (✉)
University of Rennes, Inria, CNRS, IRISA, Rennes, France
e-mail: davide.frey@inria.fr

Z. Á. Mann
University of Amsterdam, Amsterdam, Netherlands
e-mail: z.a.mann@uva.nl

A. Palomares
Atos, Madrid, Spain
e-mail: angel.palomares@atos.net

L. Piatti
InfoCert, Rome, Italy
e-mail: lorenzo.piatti@infocert.it

J. Schroers
KU Leuven, Leuven, Belgium
e-mail: Jessica.Schroers@law.kuleuven.be

N. El Madhoun et al. (eds.), *Building Cybersecurity Applications with Blockchain and Smart Contracts*, Signals and Communication Technology,
https://doi.org/10.1007/978-3-031-50733-5_6

types of DLT approaches (public or private, permissioned or permissionless). It shows that three fundamental properties of DLTs—immutability, decentralization, and automation—make it very difficult to comply with the GDPR in the public permissionless setting. For other DLTs, GDPR compliance is less problematic, but some challenges remain. In particular, the uncertainty that remains about the exact interpretation of the GDPR in the context of DLTs means that the question of GDPR compliance cannot always be definitely answered.

Keywords Distributed ledger · DLT · Blockchain · Data protection · Privacy · GDPR

1 Introduction

The importance and impact of distributed ledger technologies (DLTs) grew quickly in recent years. The success of Bitcoin [1] inspired many other cryptocurrencies, including Ethereum, Litecoin, and Ripple. Blockchain, the technology underlying Bitcoin and several other cryptocurrencies, also found many other applications beyond the field of cryptocurrencies, including accounting and auditing, regulatory reporting, supply chain management, and healthcare [2].

A blockchain is a ledger that supports appending new items, but does not allow changing or deleting items that were added to the ledger in the past. The distributed and decentralized nature of the blockchain implies that multiple parties need to agree on the content of the ledger, without needing to trust each other or a central authority. Since there are a variety of designs with these properties which significantly differ from the original blockchain, we use the more general term DLT (distributed ledger technology).

Although DLTs were introduced for cryptocurrency, their flexibility has enabled applications in a variety of contexts. For example, notaries have been looking at DLTs as a solution to digitize their profession.[1] Some projects have tried to use blockchains for the management of medical data [3]. Others like MIT's Enigma started off with a vision that blockchain would be a tool to "protect personal data" [4]. But applying DLT to personal data results in significant challenges, in terms of privacy [5] and compliance with data protection regulations.

This chapter provides a combined legal and technical analysis of data protection in DLTs. In terms of data protection regulations, we focus our attention on the General Data Protection Regulation (GDPR) of the European Union (EU) and cover all relevant provisions of the GDPR. In terms of DLTs, we look at different setups according to who can access the ledger (public vs. private, permissioned vs. permissionless DLTs). In particular, we identify three core properties of DLTs that are relevant to data protection: immutability, decentralization, and automation.

[1] https://joinup.ec.europa.eu/collection/blockchain-egov-services/solution/blockchain-based-notary-proof-concept

We analyze the challenges of GDPR compliance stemming from each of these properties.

We give here an example of the implication of each of these properties. First, as already mentioned, DLTs do not allow changing or deleting items that were added to the ledger. But the GDPR stipulates the "right to be forgotten": enabling data subjects to have their data removed from storing and processing. This clearly contradicts the immutability of DLTs [6]. Second, DLTs are based on a decentralized architecture with no central authority. Yet, the GDPR relies on the identification of a special actor, the data controller, which is responsible for ensuring compliance with the regulation. Unfortunately, it is currently unclear how a data controller should be identified in DLT-based applications [7]. Finally, several DLTs support smart contracts: software programs that can be stored on the DLT and executed in the context of the DLT in an automated way. The concept of smart contracts significantly extends the possibilities of DLTs and has contributed to their take-up [8]. But their usage in the presence of personal data raises important challenges with respect to several rights and principles stated by the GDPR, for example, the right not to be subject to solely automated processing.

Our findings indicate that these three core properties (immutability, decentralization, and automation) make it very difficult to comply with the GDPR in the public permissionless setting. For other DLTs (public permissioned, private permissioned), our analysis reveals a multifaceted situation. In these cases, GDPR compliance is less problematic, but several challenges remain. For example, our analysis of the literature reveals that there is still uncertainty about the exact interpretation of the GDPR in the context of DLTs, so that the question of GDPR compliance cannot always be clearly answered.

The remainder of this chapter starts by providing a background on DLTs and the GDPR in Sect. 2. Then, in Sect. 3, it classifies the challenges associated with the combination of DLTs, personal data, and GDPR by considering the three dimensions we suggested here: immutability, decentralization, and automation. Finally, Sect. 4 concludes the chapter.

An analysis of how different technical and legal approaches can be used to mitigate the identified challenges is given in our companion chapter [9].

2 Background

This section introduces the background that is necessary to understand the subsequent analysis. We start with the relevant technologies underlying DLTs in Sect. 2.1, followed by the legal background on data protection in Sect. 2.2. Readers familiar with the technical or legal background can safely skip the corresponding subsection. We also note that Sect. 2 in Chapter "Blockchain-Integrated Identity Verification for E-Government Services" earlier in this book [10] already provided some background on blockchain and smart contracts. Here we simply introduce some additional concepts and precisely define some terms that are relevant to our analysis.

2.1 *Technological Background*

Although the term blockchain has now become an everyday word, its key property lies in the ability to implement a ledger, i.e., an append-only list of items. For this reason, a number of authors have proposed the term "distributed ledger" and the acronym DLT for "distributed ledger technology." DLTs put together concepts and ideas from different areas of computer science. In the following, we define the main concepts and ideas employed in distributed ledger technology and discuss how they are used in this context. Throughout the discussion, we will use the term node to refer to a device that contributes to the operation of the DLT (a miner or a full node in Bitcoin terminology) and the term clients to refer to devices that people use to connect to nodes or to intermediary services such as online wallets.

We also observe that significant research is currently being devoted to alternatives to the blockchain model [11–16]. In particular, multiple research groups have shown that consensus, and thus a blockchain, is not necessary to implement a money transfer abstraction or even some types of smart contracts [15, 16]. This has led to significant work on consensus-less primitives [13, 14, 17]. In the following, nonetheless, we will concentrate on distributed ledgers that follow the classical blockchain model and establish an append-only and immutable total order among operations or blocks of operations.

2.1.1 Types of Ledgers

A first way to classify distributed ledgers stems from the environment in which they are designed to operate. In this respect, we can distinguish two types of environments: permissionless and permissioned [18].

- The **permissionless** setting consists of an environment in which devices can join and leave the system without any authorization. These systems can thus comprise an unbounded number of participants whose identities are not known a priori. The well-known Bitcoin [1] and Ethereum [19] blockchains belong to this category. Anyone can join the network and participate in the protocol at any time.
- The **permissioned** setting [20, 21], on the other hand, requires participating nodes to have the authorization of a managing authority, for example, a company or a bank. In this case, the set of participants is generally known a priori, and a change like the addition of a new participant must go through the managing authority.

The permissioned setting offers a simpler and more controlled environment, while the permissionless (also known as unpermissioned) setting opens the door to what is commonly known as Sybil attack [22], in which a node can impersonate a large number of identities (either other nodes or made-up identities), thereby influencing the outcome of the protocol.

The distinction between permissioned and permissionless refers to the ability of network nodes to participate in the DLT protocol. On the other hand, users can also interface with the ledger by operating as clients that access the DLT's data without being part of the DLT protocol. From the perspective of clients, we can thus distinguish public and private DLTs.

- In a **public** DLT, any client can access the system without any authorization and without being part of a pre-defined group.
- In a **private** DLT, clients must be registered with some managing authority and can only access the system if they have the appropriate authorization.

If all users operate as nodes, then the two classifications coincide, but in the general case, they lead to four combinations: private permissionless, public permissionless, private permissioned, and public permissioned. The private permissionless combination does not make much sense, as it would mean that clients need authorization while nodes do not. So we obtain three meaningful types of DLTs:

- **Public permissionless DLT**: public ledger with no access restrictions. Anyone with an Internet connection can send transactions, become a block validator, and participate in the execution of a consensus protocol. Examples include Bitcoin [1] and Ethereum [19].
- **Private permissioned DLT**: DLTs placing restrictions on who is allowed to participate in the network and in what transactions. They are mainly useful for business and industrial applications. In fully private permissioned DLTs, write permissions are kept centralized to one organization, while in consortium DLTs, they may comprise a consortium of organizations. Examples of private permissioned DLTs are based on platforms like Hyperledger[2] and Corda.[3]
- **Public permissioned DLT**: a new type of network that fills the gap between public permissionless and private consortium networks. A public permissioned blockchain network combines the permissioned nature of a private consortium with a decentralized governance model, trying to achieve the best properties of both models. This approach makes it possible to obtain features required for the implementation of use cases that do not fit any of the previously explained models. Alastria,[4] EOS,[5] and Ripple[6] are examples of such DLTs.

[2] https://www.hyperledger.org/

[3] https://www.corda.net/

[4] https://alastria.io/en/

[5] https://eos.io/eos-public-blockchain/

[6] https://ripple.com/

2.1.2 Inner Workings of a Ledger

A distributed ledger essentially consists of two main components: a distributed data structure and a consensus protocol that allows nodes to agree on the content of the data structure. The most common type of data structure consists of a list of blocks, commonly known as a blockchain, but also other structures exist. For example, some ledgers operate by creating DAGs (directed acyclic graphs) of blocks or transactions. But in all cases, ledgers rely on a set of basic notions in the context of cryptography and distributed systems, which are described next.

2.1.3 Cryptographic Background of DLTs

On the cryptography side, all ledgers rely on cryptographic hash functions and public key cryptography. In the following, we define these terms.

Cryptographic Hash Functions The term "hash function" generally refers to any function that takes an arbitrary size input and outputs a fixed size output. In the DLT context, the term refers to the more specific category of cryptographic hash functions. Any hash function must be deterministic (always return the same hash value for the same input), but a cryptographic hash function must also satisfy additional constraints:

1. It should be quick to compute.
2. It should be a one-way function, that is, it must be time-consuming and expensive to generate an input from its hash value.
3. It should present an avalanche effect: any small change in the input should produce big changes in its hash value.
4. It should be collision resistant: it should be unfeasible to find two different inputs with the same hash.

Property # 2 implies that the only way to go back to the original from a hash is to try all the possible variations to see if they produce a match, which is a time-consuming and very expensive task. For example, the Bitcoin network uses Secure Hash Algorithms (SHA), such as SHA-256 [23] to implement a crypto-puzzle. Property # 3 entails that if one single bit of input data is changed, the output changes significantly. For brevity, in the following, we will use the term hash function to refer to a cryptographic hash function with the above properties.

Public-Key Cryptography Public-key cryptography belongs to the family of asymmetric cryptography systems. It is based on the use of two keys, in order to overcome the limitations of symmetric crypto-systems based on a single key. In a symmetric system, participants need a secured way (e.g., a physical meeting) to agree on the key to be used for the communication, which is not feasible for modern network-based communication. Public-key cryptography solves this issue by introducing two keys: the private key and the public key. The private key is only known by the owner and needs to be kept private, while the public key should be given to anyone in the network in order to be used by anyone to send encrypted messages to the owner of the public key. Messages encrypted with the public key can only be decrypted with the corresponding private key.

Zero Knowledge Proof (ZKP) ZKP is a cryptographic technique that can be used to hide a piece of information while still making it possible to perform verification on this data. For example, a prover can use a ZKP to prove to a verifier that they own some secret data without leaking the actual content of this data. The verifier only knows the fact that the prover owns this data. For example, in the context of a cryptocurrency, a ZKP can be used to generate proof that the prover has enough funds for a transaction without revealing the exact amount it owns [24].

2.1.4 Distributed Systems Background of DLTs

From a systems' perspective, a DLT can be viewed as a distributed state machine [25, p. 313], an abstract object that creates the illusion of having a widely available unified computing system regardless of distribution or failures [26]. This has two direct implications. First, a ledger can support Turing-complete languages as done in the Ethereum cryptocurrency [19]. Second, it requires the ability to solve distributed consensus [26].

Distributed consensus models a set of participating devices that need to agree on an outcome. Each participant proposes a value, and at the end of the algorithm, each decides on a value. In the case of a ledger, the value can for example be the content of the next block that should be appended. A major result by Fischer, Lynch, and Paterson [27] states that consensus cannot be solved in the presence of failures (even benign crash failures) in an asynchronous system (i.e., where messages can experience unbounded delays). Intuitively, this results from the impossibility to distinguish a failed participant from an extremely slow one. In practical systems, this impossibility can be circumvented in several ways: e.g., assuming that communication is partially or eventually synchronous [28–30] or by employing randomized algorithms with probabilistic guarantees. This has led to the appearance of several systems that effectively exploit distributed consensus in the presence of crash [31] and Byzantine (arbitrary) failures [32–36].

Consensus in Permissionless Settings Classical distributed consensus algorithms [28–36] operate under the assumption that the distributed system consists of a predefined number of well-identified participants. One of the novelties of blockchain protocols was to operate in an open (permissionless) environment, in which devices can join and leave the system without needing to register the fact that they are joining the system. This requires solutions to withstand identity-based attacks such as the Sybil attack, in which a node can impersonate a large number of identities.

As a result, classical consensus algorithms do not fit the permissionless setting defined by public ledgers. For this reason, Bitcoin and subsequent designs have based their operation on a novel family of consensus algorithms that we can name Proof-of-*. Proof-of-* follows a well-known means to achieve distributed consensus. It elects a leader, and then it allows this leader to take one (e.g., Bitcoin, Ethereum) or more (e.g., BitcoinNG) decisions. This model was already used in the closed context classical distributed systems [36–38], but the novelty of proof-of-* consists in enabling leader election, and thus consensus, in permissionless systems in the presence of Byzantine and Sybil attacks. The very nature of proof-of-* protocols consists in a Sybil resilience mechanism that makes it hard for attackers to generate arbitrary identities.

The best known proof-of-* protocol is the proof-of-work (PoW) invented by Malkhi [39] and first used in the context of DLTs by Bitcoin's creator, Satoshi Nakamoto [1]. Given the heavy utilization of energy and computational power consumed by PoW, other consensus mechanisms have emerged in the last few years to mitigate PoW inefficiencies, such as proof-of-stake (PoS), delegated-proof-of-Stake (DPoS), Proof-of-Burn (PoB), Proof-of-Authority (PoA), Proof-of-Elapsed Time (PoET), and Proof-of-Capacity (PoC) [40].

Consensus in Permissioned Settings Even if we were to ignore its high computational cost, PoW becomes much less appealing in permissioned settings, where attacks like the balance attack [41] become easier than in permissionless ones. But even other proof-of-* solutions become less interesting. On the one hand, the main motivation for proof-of-* solutions, the possibility of Sybil attacks, disappears in a permissioned setting where all participants are known. This has led the major permissioned solutions to adopt deterministic BFT consensus protocols from the domain of distributed computing. While these cannot currently operate in large-scale systems in permissionless settings, they present two major advantages with respect to proof-of-*. First, they can achieve very high throughput. Second, they provide decision finality. In proof-of-*, a decision can be reverted with a probability that decreases with its depth in the chain. BFT protocols make decisions final as soon as they are taken and do not even need a chain to achieve consensus [36, 42].

2.1.5 Automation and Smart Contracts

Chapter "Blockchain-Integrated Identity Verification for E-Government Services" [10] already defined smart contracts. Here we simply observe that smart contracts can bring several advantages with respect to traditional contracts, such as increased security (data managed through a smart contract are immutable and tamper-proof), increased speed (all transactions are automated and quicker than a manual system), increased accuracy (smart contracts automatically follow their built-in rules, greatly reducing the potential for error and simplifying verifiability by third parties), and decreased cost (smart contracts streamline transactions and effectively remove the middleman, thereby lowering transaction cost).[7] But they can also introduce new challenges as we discuss in the following.

A DAO (decentralized autonomous organization) is a complex smart contract or a set of smart contracts. A DAO can be used to manage elaborate situations in which more parties (data subjects) are involved. DAOs usually implement decision-making systems to enable an online community to reach agreements. As a result of these agreements, the DAO operates automatically by executing the appropriate portion of code on the blockchain network [43].

2.2 *Data Protection Under the GDPR*

The European Union (EU) considers data protection to be a policy priority and a critical issue on the path of digitizing society toward a citizen-friendly single digital market [5]. The EU has adopted the General Data Protection Regulation (Regulation 679/2016, GDPR) which contains basic principles of data protection along with obligations, rights, and duties for each of the data environment stakeholders.

2.2.1 The GDPR's Scope of Application

The GDPR applies to the processing of personal data. The notion of "personal data" covers "any information relating to an identified or identifiable natural person ('data subject'); an identifiable natural person is one who can be identified, directly or indirectly, in particular by reference to an identifier such as a name, an identification number, location data, an online identifier or to one or more factors specific to the physical, physiological, genetic, mental, economic, cultural or social identity of that natural person" (art. 4 (1) GDPR).

Processing means "any operation or set of operations which is performed on personal data or on sets of personal data, whether or not by automated means" (art. 4 (2) GDPR), which includes every action done with personal data: the sole fact of starting data collection triggers the application of the regulation.

In certain cases the GDPR is not applicable, for example, when data processing is carried out by a natural person in the course of a purely personal or household activity (the household exemption) or when processing data that have been anonymized.

Anonymization of personal data means that the data is irreversibly de-identified [44]. This is different from pseudonymization or encryption, where the original data can be re-identified using appropriate additional information. Pseudonymized and encrypted data is still personal data, since it can be related to a

[7] https://blockchain.ieee.org/topics/smart-contracts-and-energy-how-blockchain-smart-contracts-can-improve-the-energy-sector

natural person with additional information, such as a decryption key. Therefore, in case of pseudonymization or encryption, the GDPR is applicable.

It is not always clear whether a person can be considered identifiable. A decisive factor in this regard is the concept "means likely reasonably to be used" [45]. There are two main approaches for interpreting this concept: the "absolute" and the "relative" approach [46]. The difference is whether the means must be available to anybody (absolute) or only to the controller (relative). For DLTs, the difference is important in the case of off-chain storing of personal data and public keys (see also the section on keeping personal data off-chain in our solutions chapter [9]):

- Following the absolute approach, the pointer on the blockchain to the off-chain data could be considered personal data since it relates to an identifiable person (identifiable via the additional information available off-chain). Since public keys can potentially be connected to information about a person, they would be considered personal data.
- Following the relative approach, it would only be personal data if the party processing the data on the blockchain is indeed able to obtain the off-chain information [46].

Currently, the direction of interpretation in practice seems to be toward the absolute approach, based upon the wording of the GDPR and the Article 29 Working Party opinions [46].

The European Court of Justice (CJEU) in the Breyer decision [45] considered that the possibility to combine a dynamic IP address with the additional data held by the Internet service provider (ISP) could constitute a means likely reasonably to be used to identify the data subject. According to the case, it is important to consider whether identification is reasonable. Since legal channels exist to obtain information from the ISPs, IP addresses are considered personal data. However, it would not be considered personal data if the identification of the data subject was prohibited by law or practically impossible due to the required disproportionate effort in time, cost, and manpower, resulting in an insignificant risk of identification. Moreover, due to the way the question to the court was phrased, it is not entirely clear whether the court follows the relative approach [46].

2.2.2 Principles of Data Processing in the GDPR

According to art. 5(1), data processing must be carried out in accordance with the following principles: lawfulness, fairness and transparency; purpose limitation; data minimization; accuracy; and integrity and confidentiality. Moreover, according to art. 5(2), the data controller is responsible for and must be able to demonstrate compliance (the accountability principle).

2.2.3 Role of the Data Controller

Central to data protection law is the notion of the data controller, described in the GDPR as "the natural or legal person, public authority, agency or other body which, alone or jointly with others, determines the purposes and means of the processing of personal data." It can do so alone or jointly with other entities (in that case we speak of joint controllers) (art. 4(7)). The data controller carries the general obligation of compliance (art. 5(2) and 24 GDPR). It may entrust a data processor to process data on the controller's behalf, without having the power to determine its means or purposes. In practice, it may be difficult to distinguish between controllers and processors due to technological and societal developments as well as the lack of legal flexibility [47]. Furthermore, it is required to take appropriate measures to facilitate the exercise of data subjects' rights and to provide data subjects with clear information about their rights (art. 12(1) and 12(2) GDPR).

2.2.4 Data Subject Rights

The GDPR gives the data subject the following rights in articles 13–22:

- Information rights (art. 13 and 14 GDPR): With regard to the transparency principle, the data subject has the right to receive information.
- Right of access (art. 15 GDPR): The data subject has the right to get confirmation from the controller whether his/her personal data are processed, and in such a case, access to the personal data and certain information, including the existence of automated decision-making/profiling, or about the existing safeguards if personal data is transferred to a third country. The right of access also specifies that the controller shall provide a copy of the processed personal data, as long as it does not adversely affect the rights and freedoms of others.
- Right to rectification (art. 16 GDPR): The data subject may obtain rectification of inaccurate personal data without undue delay. This right includes a notification obligation of the controller if the data has been transferred to other controllers, unless this proves impossible or involves disproportionate effort (art. 19 GDPR).
- Right to erasure (art. 17 GDPR): Applies under certain circumstances such as when the data is no longer necessary, the data subject withdraws consent or objects to the processing, the data have been unlawfully processed or must be erased for compliance with legislation, or if the data have been collected in relation to the offer of information society services. This right also includes a notification obligation of the controller if the data has been transferred to other controllers, unless this proves impossible or involves disproportionate effort (art. 19 GDPR).
- Right to restriction of processing (art. 18 GDPR): Under certain circumstances and for certain periods of time, the data subject can require the controller to restrict the processing of his or her personal data. If the right is employed, then again a notification obligation of the controller exists if the data has

been transferred to other controllers, unless this proves impossible or involves disproportionate effort (art. 19 GDPR).

- Right to data portability (art. 20 GDPR): The right to data portability enables the data subject to transfer data between different controllers. It requires the controller to provide the data in a structured, commonly used and machine-readable format and that the data subject has the right to transmit those data to another controller without hindrance from the original controller and where technically feasible and requested by the data subject, the personal data should be directly transmitted from one controller to another.
- Right to object (art. 21 GDPR): The data subject has the right to object against the processing of his or her personal data on grounds relating to his or her particular situation. This right can only be invoked by the data subject if the processing is based upon public interest or the legitimate interest of the controller or processed for direct marketing purposes.
- Right not to be subject to a decision based solely on automated processing (art. 22 GDPR): The data subject has the right that decisions that produce legal effects or similarly significantly affect him or her should not be solely based on automated processing, including profiling. There are certain exceptions to this right, e.g., if it is necessary for entering into, or performance of, a contract between the data subject and a data controller; if it is authorized by Union or Member State law to which the controller is subject and which also lays down suitable measures to safeguard the data subject's rights and freedoms and legitimate interests; or if it is based on the data subject's explicit consent.

2.2.5 Transfer of Personal Data to Other Jurisdictions

Many organizations that store or process personal data are located outside the EEA (EU, Norway, Iceland, and Liechtenstein). Such data transfers are only compliant if the level of protection guaranteed by the GDPR is not undermined (art. 44 GDPR) [48]. From a legal perspective, the GDPR defines four options for establishing that a non-EEA country satisfies the required standards of data protection: (1) adequacy decisions (art. 45 GDPR), (2) appropriate safeguards (art. 46 GDPR), (3) binding corporate rules (art. 47 GDPR), and (4) exception-explicit consent (art. 49 GDPR).

- Adequacy decision: only allowing transfers to countries which the Commission considers adequate in their level of protection (art. 45 GDPR, examples are the UK, Israel, and South Korea).
- Appropriate safeguards: if appropriate safeguards are ensured (art. 46 GDPR, meaning inter alia adopting measures preventing access by national security agencies).
- Binding corporate rules: if the transfer is within the same group of undertakings, subject to binding corporate rules and having been confirmed by a supervisory authority (art. 47 GDPR).

– Exception-explicit consent: if the data subject has given explicit consent after being informed about the possible risks. A transfer under this exemption is only possible if not prohibited by EU or national law (art. 49 GDPR).

A large number of countries do not fall into any of these categories. For example, the USA, which, in the case of Bitcoin, comprises 16% of all the identified nodes,[8] is not considered by the European Commission to have an adequate level of protection. Until 2020, US-based data controllers relied on the privacy shield agreement, through which they were able to comply with some essential data protection requirements. However, the agreement was invalidated by the Court of Justice in its Schrems II decision [49]. The same decision also invalidated the Commission-approved standard contractual clauses. Although a new regime is being discussed on the political level, controllers are now left with the option of transferring data based on binding corporate rules within their own group of undertaking or appropriate safeguards under art. 47 of the GDPR. Both of these appear difficult to apply in a blockchain context.

3 Data Protection Challenges of DLTs

This section gives an overview of the main challenges in data protection in connection with DLTs. DLTs have some intrinsic features that, on the one hand, make them suitable for specific purposes and use cases but, on the other hand, may give rise to some potentially severe data protection issues. The key properties of DLTs that make data protection challenging are **immutability**, **decentralization**, and **automation**. We start by reviewing the challenges stemming from these properties and then provide some practical examples of the challenges. Table 1 provides a visual summary of the issues discussed in this section.

3.1 Challenges Resulting from the Immutability of DLTs

Immutability constitutes one of the pillars of DLT solutions. In the context of DLTs, immutability refers to the fact that information in old-enough blocks cannot be changed [50]. This constitutes a challenge from a data protection point of view, in particular for data subject rights and the implementation of data protection principles. Since the data stored in a block normally cannot be modified afterward, it means that it is not possible to comply with certain data subject rights, in particular the **right to rectification** (art. 16 GDPR) and the **right to erasure** (art. 17 GDPR), since the data cannot be changed or erased at request. Furthermore, the immutability

[8] https://bitnodes.io/

Table 1 Summary of challenges

	Public permissioned	Public permissionless	Private permissioned
IMMUTABILITY			
1st challenge: Irreversibility of DLT ⇒ challenges for data subject rights	Possibility to change content, ledger not immutable, depends on consensus mechanism & number of nodes. 0 +	Very challenging to comply with data protection rules. + +	Possibility to change content, ledger not immutable, depends on consensus mechanism & number of nodes. 0 +
DECENTRALIZATION			
2nd challenge: Identification of Controllers and Processors	Nodes are identified and authorized to create the ledger, data protection rules are enforceable. 0	Nodes not identified nor authorized to create the ledger, data protection rules are NOT enforceable. + +	Nodes are identified and authorized to create the ledger, data protection rules are enforceable. 0
3rd challenge: Transfer of data outside the EU	Restrictions on location can be implemented, data protection rules are enforceable. 0	No clear solutions for restricting node location, data protection rules are NOT enforceable. + +	Restrictions on location can be implemented, data protection rules are enforceable. 0
4th challenge: Consent management in a decentralized environment	Reading non-authorized ⇒ difficult to design correct consent management procedure. +	Reading non-authorized ⇒ difficult to design correct consent management procedure. +	As reading is authorized, consent and privacy notice can be managed. 0
AUTOMATION			
5th challenge: Automation of decision made with personal data,	Challenging to implement a correct data protection approach. +	Challenging to implement a correct data protection approach. +	Solvable with the correct data protection approach (consent or other legal basis). 0

0 **Not a problem**: due to the technological characteristics of the given type of DLT, the challenge does not pose a problem.

\+ **Issue**: the challenge does pose an issue, but it can be easily solved with a legal or a technical solution, without distorting the DLT approach.

\+ + **Big issue**: the challenge does pose an issue, which cannot be easily solved, neither with a legal nor with a technical work-around.

arises from the regular verification that the hashes of the data on the chain are still correct, which means that the data on the chain is processed. This gives rise to some uncertainty regarding the **right to restriction of processing** (art. 18 GDPR) and the **right to object** (art. 21 GDPR). With respect to the former, it has not yet been clearly decided whether verifying the content and hashes of a blockchain would mean an infringement of the restriction of processing or whether this could be considered functional to storage, and therefore it would fall under the storage exemption. The **right to object** (art. 21 GDPR) does not have such a storage exemption, but the fact that processing is functional to the operation of the DLT could constitute grounds to invoke the legitimate interest of the controller to continue using the DLT. However, this would have to be evaluated on a case-by-case basis with a careful balancing analysis.

The immutability of DLTs also threatens data protection principles, which may overlap with data subject rights. The inability to rectify incorrect data makes it impossible to comply with the **accuracy principle** (art. 5 (d) GDPR). Similarly, the inability to remove or anonymize data clashes with the **principle of storage limitation** (art. 5 (e) GDPR): data can legally be stored at length only for archival purposes in the public interest, scientific or historical research purposes, which is not the case in many DLT use cases. The **principle of purpose limitation** (art. 5 (b) GDPR) requires that personal data be collected only for specified, explicit, and legitimate purposes and not further processed for a different purpose, except if it can be argued that the processing is compatible with the original purpose. The challenge is establishing whether processing on a DLT is really needed for the purpose at hand. This can only be dealt with on a case-by-case basis. Similarly, the **principle of data minimization** (art. 5 (c) GDPR) requires that data should be adequate, relevant, and limited to what is necessary in relation to the purposes for which they are processed. Also, this principle could be breached if the data is still further processed in the DLT: since the data cannot be deleted, it is known from the start that in the end more data would be processed than necessary. Immutability also poses problems if **consent** is used as legal ground for processing, as the data subject has the right to withdraw consent at any time.

Moreover, the GDPR requires the implementation of **data protection by design and by default** (art. 25 GDPR). In a DLT, this means that the controller, when deciding on the technology to be used for processing, but also during the processing, should implement appropriate technical and organizational measures to implement data protection principles and to integrate the necessary safeguards into the processing to meet the requirements of the GDPR and protect the data subject rights. Furthermore, the controller must make sure that by default, only personal data which are necessary for each specific purpose of the processing are processed. However, if the controller implements a DLT with the safeguards that seem sufficient today, but later the safeguards become insufficient, then the immutability of the DLT may prohibit the controller from moving to another solution.

3.2 Challenges Resulting from the Decentralization of DLTs

The decentralized nature of DLTs leads to two categories of challenges. First, the potentially large number of users and devices that can constitute a blockchain makes it particularly hard to identify the roles they should take in the application of the GDPR. Second, the geographical spread of these nodes, together with the replication strategies adopted by DLTs, causes data to be stored and processed across diverse jurisdictions, making compliance with the GDPR particularly difficult.

Several principles and rights laid out in the GDPR, starting with, but not limited to, the **principle of accountability** of art. 5(2) and **general responsibility** under art. 24, rely on the identification of a party as the data controller. However, identifying roles (data controller, data subject, and data processor) constitutes a fundamental challenge in a DLT that handles personal data, since data subject rights are at stake if a controller cannot be identified [51]. The decentralized nature of distributed ledgers makes it impossible to identify these roles without ending up with absurd interpretations. For example, [7] discusses how the decentralized and peer-to-peer nature of permissionless ledgers only allows two solutions for the identification of the controller: either each node is a controller, or there is no controller. Both of these solutions would result in the inapplicability of GDPR to the blockchain.

A more reasonable stance with respect to this challenge is put forward by Moerel, who observes that DLTs constitute a form of general-purpose technology, like the Internet itself [52]. Therefore, the GDPR should not be directly applied to DLTs, but to the applications built on top of DLTs. For example, a company or user that uses a smart contract to process personal data would become the data controller, the nodes running the blockchain, and thus executing the contract would be at most data processors following the definition under art. 4(8). In this respect, ledger nodes can be compared to nodes in the cloud that store data on behalf of some client of the cloud platform [53].

The other challenge related to distribution results from the geographical spread of blockchain nodes, which may be located anywhere in the world. The GDPR requires that **transfers to third countries** can only take place insofar **an adequate level of protection** of personal data can be guaranteed (art. 45) or other guarantees are put in place. Since most blockchain platforms adopt a total replication strategy, i.e., they replicate each block on all the nodes that make up the system, this raises not only scalability issues, but it also implies that data will probably be stored outside the EU, possibly in countries that do not provide an adequate level of data protection guarantees. This problem presents significant challenges, particularly in the context of permissionless blockchains. A public permissionless blockchain may involve nodes from any country, and, at least in currently deployed systems, no one can have control on where nodes, let alone clients, are located. This makes permissionless blockchains the most challenging scenario from the point of view of the tension between GDPR and decentralization.

Paradoxically, while determining a controller on blockchains can be challenging, sometimes blockchains can instead lead to easier designation of a controller in some instances [54]. This is due to the persistence and record-keeping functions of the blockchains—shipping companies such as Maersk use blockchains to enhance transparency in their supply chains.

Lastly, true decentralization involves the possibility of global use of blockchain technology: this means that a user—data subject in the GDPR semantics—can access services at any time and without any control. It becomes, therefore, difficult to show privacy notice, in compliance with Article 12, or—if necessary—to collect consent to processing, in compliance with Articles 6(1)(a) and 7.

3.3 Challenges Resulting from the Automation in DLTs

Automation—as described in Sect. 2.1.5—is relevant to the GDPR for three main reasons. First, because the regulation sets a specific perimeter for automation in data processing. Second, because automation can be a tool for unlawful processing by third party or external code invocation. Third, automation may create some friction with certain GDPR principles, as explained below.

Regarding the first issue—specific perimeter—the European regulator focuses on three points:

- *Automatic decision-making.* According to article 22, the data subject has the right not to be subject to a decision based solely on automated processing, which produces legal effects concerning him or her or similarly significantly affects him or her. In a DLT environment, there are two types of applications that can potentially threaten this right: smart contracts and decentralized autonomous organizations (DAOs). Using smart contracts and/or DAO solutions obliges the data controller—assuming that a data controller can be identified—to implement suitable measures to safeguard the data subject's rights and freedoms and legitimate interests, at least the right to obtain human intervention on the part of the data controller, to express the data subject's point of view and to contest the decision. Also, data controller shall then communicate in an accurate way how the smart contract itself works, according to Recital 63 "Every data subject should therefore have the right to know and obtain communication in particular with regard to the purposes for which the personal data are processed, where possible the period for which the personal data are processed, the recipients of the personal data, the logic involved in any automatic personal data processing and, at least when based on profiling, the consequences of such processing."
- *Profiling.* The definition of profiling is also based on the concept of automation, as—per recital 71—it is "any form of automated processing of personal data consisting of the use of personal data to evaluate certain personal aspects relating to a natural person." Under this provision, the data controller has the burden to grant a correct and specific legal basis for the processing.

- *Portability.* Automation is also relevant under article 20 (1) (b), which grants to the data subject the right to portability when "the processing is carried out by automated means." According to this provision, data controller shall grant the data portability each time the processing is made by tools—in our case smart contracts/DAOs—that enable automated processing.

Regarding the second issue—unlawful processing—automation can be misused in a blockchain environment. Open access and open use of DLTs may make GDPR compliance difficult: third and unknown parties, which may not be allowed for a specific data processing, may use the related data. For example, in a scenario where a specific set of data is used within automation processes without authorization criteria, the vast number of subjects who can—by exploiting the automatism and the functioning of smart contracts—access personal data could make it difficult to reach a GDPR-compliant approach. A specific example could be the misuse of a database with the records of a decentralized gaming platform: the operation of the platform would have a very high degree of automation, and access to data could be indiscriminate in the absence of precise rules in this regard.

Regarding the third issue—friction with certain GDPR principles—automation in DLTs may affect further general rules and duties as follows:

- *Principle of lawfulness, fairness, and transparency:* automation by smart contracts may violate this principle, as the data subject could be faced with a computer object that is not intelligible and therefore violates the principle of transparency. The same could apply also to lawfulness and fairness, given that the tool used may not reflect all the information necessary for the data subject to be aware of how their data is being processed, resulting in a non-compliance of article 5(1)(a).
- *Principle of confidentiality:* smart contracts create a mechanism whereby access to data—in the event of incorrect, or malicious, implementations—may be indiscriminate, violating the principle in question.
- *Controller shall maintain a record of processing activities under its responsibility (art. 30 GDPR):* in a highly automated context, it is difficult to analyze and identify each treatment. Smart contracts enable massive treatments, often independent of the will of the individual. Without proper tools, it may be difficult to keep track of individual operations, despite the transparency of the blockchain.
- *Controller shall notify personal data breach (art. 33):* identifying and reporting data breaches could be difficult in a context where data access is available to software that automates processing, e.g., smart contracts. If there are no precise authorization rules, it may be hard to identify when and how an access has caused a data breach. Moreover, in its broadest sense, the data breach includes the impossibility to access the data, and a process exploiting DLT automation may lead to a situation in which a software malfunction prevents the use of some personal data.

As an example of the above three issues, we can imagine a scenario where a seller (acting as a data controller) of digital goods or services may use a smart contract that automatically processes data about a buyer (data subject). The seller uses the buyer's data to create, complete, and deploy the smart contract to manage the sale and the execution of the contract between the parties. These data, once entered into the blockchain, will be used automatically, by both the seller and the buyer, for the purpose set out in the contract, but also by other actors, who can call the same contract for related purposes.

Another example could arise in connection with DAOs. A DAO (decentralized autonomous organization) is a complex smart contract or a set of smart contracts. A DAO can be used to manage elaborate situations in which more parties (data subjects) are involved. A DAO is used to autonomously manage different rights and functions of the "organization," such as voting and sharing. DAOs, like smart contracts, may pose a threat to personal data under two circumstances. On the one hand, they can access personal data on databases, making themselves a tool for data processing. On the other hand, they receive information on how users of the DAO behave and interact, creating a potential profile of the data subjects. Such an organization may—under the abovementioned circumstances—process a huge amount of personal data, exacerbating the problems of smart contracts. Each participant gets one or more tokens which can be used to express preferences on a certain topic, for example, to automate crowdfunding or complex company actions and decisions. The movement of tokens and related decisions can be tracked on the ledger, potentially exposing personal data regarding choices or other actions and relationships of the data subjects.

A third example of DLT automation arises from cross-chain smart contracts [55]. Cross-chain smart contracts are decentralized applications composed of multiple smart contracts deployed across different DLTs that interoperate to create a single application. Decentralized services built on a single chain are often inefficient for enterprise applications that generate millions of transactions per second. Cross-chain approaches can increase computational efficiency, but also aggravate issues regarding the protection of data in a cross-chain application with intensive use of automated smart contracts. These issues fall within the area of interoperability, which is an emerging trend in DLT research, and require solutions to preserve data privacy across different blockchain networks when interactions between multiple smart contracts happen.

3.4 *Practical Examples*

We now consider three practical examples that further highlight the challenges resulting from the use of DLTs for the storage or management of personal data. We start by considering the case of public keys (Sect. 3.4.1) which pose challenges even if they constitute an integral part of the DLT ecosystem. Then we discuss managed wallets (Sect. 3.4.2), and finally (Sect. 3.4.3), we consider one of the current killer applications of DLTs: self-sovereign identity systems.

3.4.1 Public Keys

Public keys are often used in the context of DLTs, and their relevance to data protection represents an interesting dilemma. Since it is often possible to connect public keys—with additional information—to natural persons, thus allowing the identification of a natural person, public keys can generally not be considered anonymous, but rather pseudonymous [7, pp. 13–16] [56, p. 40] [57, p. 95].

Whether a public key is personal data or not may depend on information that is not directly visible. A public key consists of a number and is often represented by a string corresponding to the hexadecimal encoding of the number. In general, it is impossible to determine, by looking at a public key, whether it refers to a natural person. Even when it does, the additional data that makes this association concrete may not be available to any third parties. In some cases, there may be a company that serves as an interface between a person and the blockchain system, and that company records a link between this person's public key and his/her identity or other personal information. In other cases, this additional data may only exist on devices that are entirely under the control of the person to whom the public key refers. This is for example the case for a user that connects to a blockchain platform from his/her own computer using a client program or using a full-fledged blockchain node.

The French data protection authority, CNIL, considers public keys as personal data that are essential for the proper functioning of the blockchain [58]. But in general, there is no clear agreement in the literature as to the status of public keys as personal data. Nonetheless, since they may represent personal data, the challenges discussed in Sects. 3.1, 3.2, and 3.3 may apply to public keys.

Let us consider immutability; data subjects whose public key is processed in the DLT will not be able to use their data subject rights, such as the right to rectification and the right to erasure. The public key will stay in the DLT, and the principle of storage limitation or data minimization cannot be complied with. As it is not possible to comply with data protection principles and data subject rights, the use of DLTs with public keys would mean that the principle of data protection by design and by default would not be complied with. Considering the decentralized nature of DLTs, it would also not be possible to identify who is responsible (i.e., the controller) for the processing of the public keys, and the public key might be transferred to third countries. Similarly, in most cases, it is difficult if not impossible to ensure the art. 13 right of information with respect to processing of public keys. It is, in fact, difficult to provide the information about the processing to the data subjects, if only their public key is known.

Finally, a further issue, related to public keys, and to any other identifier, is that they can enable linkability [59] between different records on the blockchain. For example, the ability to determine the amount in a person's Bitcoin wallet relies on the fact that this person uses the same public key for all their transactions.

3.4.2 Wallets and Addresses

Blockchain is a public ledger where transaction information is stored. Each transaction happens between addresses. Wallets are a way to process and manage addresses. As long as the identity of the natural person behind an address or a wallet is not disclosed, there is no trace of personal data. However, ensuring that an address or a wallet cannot be linked to a natural person is challenging. Attackers may correlate different transactions involving the same address or may combine on-chain transaction history information with publicly available or easily obtainable off-chain information to break the anonymity of addresses or wallets [60]. Users' wallet address are public, and, although blockchain network are pseudonymous, they can be tracked, potentially revealing transaction patterns and user behavior [61] and posing a significant challenge for blockchain wallet users who desire anonymity. Another potential issue is related to private key management, where the security of the wallet depends on the safeguarding of the private key associated with the user's public address [62]. An inadequate storage and management of private keys can expose the wallet to theft, unauthorized access, or loss of funds. This is particularly true when wallets rely on third-party services to provide additional features such as data storage, exchange services, or multi-signature functionality [63]. Integrating with these services can introduce new security vulnerabilities and data protection risks if the third-party providers do not adhere to strict security standards.

3.4.3 Self-Sovereign Identity

Self-sovereign identity (SSI) has emerged in recent years as a novel paradigm that seeks to improve overcentralized or federated identity management systems. Essentially, SSI seeks to put users back at the center of the management of their digital identities by making them not only user-centric, but also completely decentralized.

Typically, SSIs are built on the concept of anonymous credentials. An anonymous credential (AC) consists of a cryptographic primitive that makes it possible for a user to prove some aspects of their identity without disclosing any information other than what they want to prove. In other words, any participant in the SSI can issue a credential to any other or verify someone else's credential. Essentially, an SSI system consists of three roles: users, issuer, and verifier. A user can be any person or device who wishes to access a service. An issuer is any entity that issues credentials to users. A verifier is whoever verifies the validity of credentials. As suggested above, a single person or entity may assume any number of roles.

A recent series of blogposts [64, 65] by Bilgesu Sümer examines the similarities and differences between the GDPR and the concept of SSI. It is not within the scope of this chapter to report all the points discussed by these blog posts; nonetheless, a few of them are worth mentioning.

In particular, the posts highlight that while the two share common concepts and ideas, they also exhibit important differences. For example, one seeks to protect the right to identity, while the other the right to privacy. Moreover, some terms assume different meanings depending on whether one is referring to SSI or the GDPR. In particular, transparency and access refer to public access to transparent data for SSIs, while it refers to a more restrictive notion in the GDPR.

According to Sümer, the main hurdle in reconciling SSI and GDPR remains associated with the persistence and immutability features of the underlying blockchain [65]. Sümer observes that the principle of data minimization requires that data be kept for as little as necessary for the required purpose and argues that the verification of a credential typically requires one second. However, to the best of our knowledge, existing SSIs do not store credentials on the blockchain. Rather, they only use a blockchain only to store information that has been chosen to be publicly available (e.g., a public DID referring to a publicly available service provider).

Yet, even if the blockchain stores only public DIDs, this still poses problems with respect to the protection of personal data. Consider a user, Bob, who opens a business that requires the verification of a person's age. Bob calls the business AnonACME, an anonymous-looking name because he does not want to be publicly associated with it. The company's DID that goes on the blockchain therefore only refers to AnonACME and not to Bob. However, Bob still needs to register the business with the public company registry, which has a publicly available database that lists Bob as the owner of AnonACME. This makes AnonACME sensitive personal data, as the public registry makes it easy to associate it with Bob.

If Bob runs his business for all his life or is otherwise proud of his business, everything looks fine. But let's imagine that Bob's runs into trouble with his business and wishes to delete it from the blockchain. All the blockchain can do is revoke its business by adding a revocation entry, but information about the previous existence of AnonACME cannot be erased. It is true that this information is also present in the public company database, but the government has legal grounds for maintaining this information, while it is not clear that such legal grounds exist for the SSI's blockchain. The persistence of the data in the blockchain can therefore be a problem even for public DIDs contradicting the principle of minimization and the right to erasure.

Although not strictly related to SSIs, another blogpost [66] analyzes the specific case of the eIDAS Regulation (a proposal by the European Union for a European Digital Identity). The authors observe that the use of persistent identifiers—implied by art. 12(4)(d) of eIDAS2[9]—directly clashes with rights and principles of GDPR, such as data minimization or the right to erasure.

[9] https://digital-strategy.ec.europa.eu/en/policies/eidas-regulation

4 Conclusions and Outlook

This chapter provided a thorough analysis of the challenges associated with the use of DLTs for the management of personal data. Our analysis concerned both legal and technical aspects, to discuss how different technical approaches to DLTs may or may not comply with the GDPR.

We identified three fundamental properties of DLTs that make compliance with—different provisions of—the GDPR challenging: immutability, decentralization, and automation. These properties create a significant problem for public permissionless DLTs. For other types of DLTs, achieving compliance with the GDPR is less difficult, but still not trivial as compliance may depend on many factors, such as the type of data stored in the DLT, the type of processing that is performed by the DLT, links to data outside the DLT, etc. We also observed that several questions about the interpretation of the GDPR in the context of DLTs are still not completely clear, for example, determining the controller for a DLT application or identifying whether public keys and hashes should be considered personal data.

These challenges make it particularly difficult to develop and operate a DLT-based application in a GDPR-compliant way while handling personal data. In our companion chapter [9], we analyze how different technical and legal approaches can be used to mitigate some of these challenges and outline directions for future research.

References

1. Nakamoto, S.: Bitcoin: a peer-to-peer electronic cash system (2008). https://git.dhimmel.com/bitcoin-whitepaper/
2. Berdik, D., Otoum, S., Schmidt, N., Porter, D., Jararweh, Y.: A survey on blockchain for information systems management and security. Inf. Process. Manag. **58**(1), 102397 (2021)
3. Panetta, R., Cristofaro, L.: A closer look at the EU-funded My Health My Data project. In: Digital Health Legal pp. 10–11 (2017). https://doi.org/10.5281/zenodo.1048999
4. Zyskind, G., Nathan, O., Pentland, A.S.: Decentralizing privacy: using blockchain to protect personal data. In: 2015 IEEE Security and Privacy Workshops. pp. 180–184 (2015). https://doi.org/10.1109/SPW.2015.27
5. Timan, T., Mann, Z.: Data protection in the era of artificial intelligence: trends, existing solutions and recommendations for privacy-preserving technologies. In: The Elements of Big Data Value: Foundations of the Research and Innovation Ecosystem, pp. 153–175. Springer, Heidelberg, Germany (2021)
6. Bayle, A., Koscina, M., Manset, D., Perez-Kempner, O.: When blockchain meets the right to be forgotten: technology versus law in the healthcare industry. In: 2018 IEEE/WIC/ACM International Conference on Web Intelligence (WI). pp. 788–792. IEEE (2018)
7. Finck, M.: Blockchain and data protection in the European Union. Max Planck Institute for Innovation & Competition Research Paper No. 18-01 (2017)
8. Hewa, T., Ylianttila, M., Liyanage, M.: Survey on blockchain based smart contracts: applications, opportunities and challenges. J. Netw. Comput. Appl. **177**, art. 102857 (2021)

9. Fabčič Povše, D., Favenza, A., Frey, D., Mann, Z.Á., Palomares, A., Piatti, L., Schroers, J.: Solutions to data protection challenges in distributed ledger and blockchain technologies: a combined legal and technical approach. In: El Madhoun, N., Dionysiou, I., Bertin, E. (eds.) Building Cybersecurity Applications with Blockchain Technology and Smart Contracts. Springer, Heidelberg, Germany (2024)
10. Talukder, S., Alam, M., Hossain, I., Puppala, S.: Blockchain-integrated identity verification for e-government services. In: El Madhoun, N., Dionysiou, I., Bertin, E. (eds.) Building Cybersecurity Applications with Blockchain Technology and Smart Contracts. Springer, Heidelberg, Germany (2024)
11. Auvolat, A., Frey, D., Raynal, M., Taïani, F.: Money transfer made simple: a specification, a generic algorithm, and its proof. Bull. EATCS **132**, 21–43 (2020)
12. Guerraoui, R., Kuznetsov, P., Monti, M., Pavlovič, M., Seredinschi, D.A.: The consensus number of a cryptocurrency. Distrib. Comput. **35**, 1–15 (2022)
13. Albouy, T., Frey, D., Raynal, M., Taïani, F.: Byzantine-tolerant reliable broadcast in the presence of silent churn. In: SSS'21. pp. 21–33 (2021)
14. Albouy, T., Frey, D., Raynal, M., Taïani, F.: A modular approach to construct signature-free BRB algorithms under a message adversary. In: OPODIS'22. vol. 253, pp. 26:1–26:23 (2022)
15. Alpos, O., Cachin, C., Marson, G.A., Zanolini, L.: On the synchronization power of token smart contracts. In: IEEE ICDCS'21. pp. 640–651 (2021)
16. Frey, D., Gestin, M., Raynal, M.: The synchronization power (consensus number) of access-control objects: the case of allowlist and denylist (2023)
17. Guerraoui, R., Kuznetsov, P., Monti, M., Pavlovič, M., Seredinschi, D.A.: Scalable byzantine reliable broadcast. In: DISC'19. vol. 146, pp. 22:1–22:16 (2019)
18. Swanson, T.: Consensus-as-a-service: a brief report on the emergence of permissioned, distributed ledger systems (2015). https://www.the-blockchain.com/wp-content/uploads/2016/04/Permissioned-distributed-ledgers.pdf
19. Wood, G.: Ethereum: A secure decentralised generalised transaction ledger. Ethereum project yellow paper (2014)
20. Monrat, A.A., Schelén, O., Andersson, K.: Performance evaluation of permissioned blockchain platforms. In: IEEE Asia-Pacific Conference on Computer Science and Data Engineering (CSDE). IEEE (2020)
21. Dabbagh, M., Choo, K.K.R., Beheshti, A., Tahir, M., Safa, N.S.: A survey of empirical performance evaluation of permissioned blockchain platforms: Challenges and opportunities. Comput. Secur. **100**, art. 102078 (2021)
22. Douceur, J.R.: The Sybil attack. In: Peer-to-Peer Systems: First International Workshop. pp. 251–260. Springer (2002)
23. Sobti, R., Geetha, G.: Cryptographic hash functions: a review. Int. J. Comput. Sci. Issues **9**(2), 461–479 (2012)
24. Morais, E., Koens, T., Van Wijk, C., Koren, A.: A survey on zero knowledge range proofs and applications. SN Appl. Sci. **1**, art. 946 (2019)
25. Raynal, M.: Fault-tolerant message-passing distributed systems: an algorithmic approach. Springer, Heidelberg, Germany (2018)
26. Lamport, L.: Time, clocks, and the ordering of events in a distributed system. Commun. ACM **21**(7), 558–565 (1978)
27. Fischer, M.J., Lynch, N.A., Paterson, M.S.: Impossibility of distributed consensus with one faulty process. J. ACM **32**(2), 374–382 (1985)
28. Dwork, C., Lynch, N., Stockmeyer, L.: Consensus in the presence of partial synchrony. J. ACM **35**(2), 288–323 (1988)
29. Dutta, P., Guerraoui, R., Lamport, L.: How fast can eventual synchrony lead to consensus? In: International Conference on Dependable Systems and Networks (DSN'05). pp. 22–27. IEEE (2005)
30. Lamport, L.: The part-time parliament. ACM Trans. Comput. Syst. **16**(2), 133–169 (1998)
31. Hunt, P., Konar, M., Junqueira, F.P., Reed, B.: ZooKeeper: wait-free coordination for internet-scale systems. In: USENIX Annual Technical Conference (2010)

32. Lamport, L., Shostak, R., Pease, M.: The Byzantine generals problem. ACM Trans. Program. Lang. Syst. **4**(3), 382–401 (1982)
33. Veronese, G.S., Correia, M., Bessani, A.N., Lung, L.C., Verissimo, P.: Efficient Byzantine fault-tolerance. IEEE Trans. Comput. **62**(1), 16–30 (2013)
34. Aublin, P.L., Guerraoui, R., Knežević, N., Quéma, V., Vukolić, M.: The next 700 bft protocols. ACM Trans. Comput. Syst. **32**(4) (2015). https://doi.org/10.1145/2658994
35. Kotla, R., Alvisi, L., Dahlin, M., Clement, A., Wong, E.: Zyzzyva: Speculative byzantine fault tolerance. ACM Trans. Comput. Syst. **27**(4) (2010). https://doi.org/10.1145/1658357.1658358
36. Castro, M., Liskov, B.: Practical byzantine fault tolerance. In: Proceedings of the Third Symposium on Operating Systems Design and Implementation. OSDI '99, pp. 173–186. USENIX Association, USA (1999)
37. Lamport, L.: The part-time parliament. ACM TCS'98 **16**, 133–169 (1998)
38. Lamport, L.: Paxos made simple, fast, and byzantine. In: Bui, A., Fouchal, H. (eds.) OPODIS. Studia Informatica Universalis, vol. 3, pp. 7–9. Suger, Saint-Denis, rue Catulienne, France (2002). http://dblp.uni-trier.de/db/conf/opodis/opodis02.html#Lamport02
39. Franklin, M.K., Malkhi, D.: Auditable metering with lightweight security. In: International Conference on Financial Cryptography. pp. 151–160 (1997)
40. Zhang, S., Lee, J.H.: Analysis of the main consensus protocols of blockchain. ICT Express **6**(2), 93–97 (2020)
41. Natoli, C., Gramoli, V.: The balance attack or why forkable blockchains are ill-suited for consortium. In: 47th Annual IEEE/IFIP International Conference on Dependable Systems and Networks (DSN). pp. 579–590. IEEE (2017)
42. Singh, A., Kumar, G., Saha, R., Conti, M., Alazab, M., Thomas, R.: A survey and taxonomy of consensus protocols for blockchains. J. Syst. Arch. **127**, Art. 102503 (2022)
43. Faqir-Rhazoui, Y., Arroyo, J., Hassan, S.: A comparative analysis of the platforms for decentralized autonomous organizations in the Ethereum blockchain. J. Internet Serv. Appl. **12**(1), 1–20 (2021)
44. Article 29 Data Protection Working Party: Opinion 05/2014 on anonymisation techniques (2014). https://ec.europa.eu/justice/article-29/documentation/opinion-recommendation/files/2014/wp216_en.pdf
45. CJEU: Patrick Breyer vs. Bundesrepublik Deutschland. ECLI:EU:C:2016:779/C-582/14 (2016). https://curia.europa.eu/juris/liste.jsf?num=C-582/14
46. Laan, V., Rutjes, A.: Privacy-issues bij blockchain: hoe voorkom of minimaliseer je die? Computerrecht **6**, 253–263 (2017)
47. Van Alsenoy, B.: Regulating data protection: the allocation of responsibility and risk among actors involved in personal data processing. Ph.D. thesis, KU Leuven (2016)
48. Schoenen, S., Mann, Z.Á., Metzger, A.: Using risk patterns to identify violations of data protection policies in cloud systems. In: Service-Oriented Computing – ICSOC 2017 Workshops. pp. 296–307. Springer (2018)
49. CJEU: Data Protection Commissioner vs Facebook Ireland Limited, Maximillian Schrems. ECLI:EU:C:2020:559/C-311/18 (2020). https://curia.europa.eu/juris/liste.jsf?language=de&num=C-311/18
50. Politou, E., Casino, F., Alepis, E., Patsakis, C.: Blockchain mutability: challenges and proposed solutions. IEEE Trans. Emerg. Top. Comput. **9**(4), 1972–1986 (2021)
51. Berberich, M., Steiner, M.: Blockchain technology and the GDPR – how to reconcile privacy and distributed ledgers. Eur. Data Protection Law Rev. **2**(3), 422–426 (2016)
52. Moerel, L.: Blockchain & data protection... and why they are not on a collision course. Eur. Rev. Private Law **26**(6), 825–851 (2018)
53. Palm, A., Mann, Z.Á., Metzger, A.: Modeling data protection vulnerabilities of cloud systems using risk patterns. In: Proceedings of the 10th System Analysis and Modeling Conference (SAM). pp. 1–19. Springer (2018)
54. IBM Security: Blockchain and GDPR: How blockchain could address five areas associated with GDPR compliance (2018). https://iapp.org/resources/article/blockchain-and-gdpr/

55. Zhang, L., Hang, L., Jin, W., Kim, D.: Interoperable multi-blockchain platform based on integrated REST APIs for reliable tourism management. Electronics **10**(23), Art. 2990 (2021)
56. Bacon, J., Michels, J.D., Millard, C., Singh, J.: Blockchain demystified. Queen Mary University of London, School of Law Legal Studies Research Paper no. 268 (2017)
57. Pesch, P., Böhme, R.: Datenschutz trotz öffentlicher Blockchain? Datenschutz und Datensicherheit **41**(2), 93–98 (2017)
58. Commission Nationale Informatique & Libertés: Blockchain – solutions for a responsible use of the blockchain in the context of personal data (2018). https://www.cnil.fr/sites/default/files/atoms/files/blockchain_en.pdf
59. Chaum, D.L.: Untraceable electronic mail, return addresses, and digital pseudonyms. Commun. ACM **24**(2), 84–90 (1981). https://doi.org/10.1145/358549.358563
60. Andola, N., Yadav, V.K., Venkatesan, S., Verma, S., et al.: Anonymity on blockchain based e-cash protocols—a survey. Comput. Sci. Rev. **40**, art. 100394 (2021)
61. Biryukov, Tikhomirov, G.: Privacy and linkability of mining pool payments. In: IEEE Conference on Communications and Network Security (CNS) pp. 118–123 (2019)
62. Tschorsch, S.: Bitcoin and beyond: A technical survey on decentralized digital currencies. IEEE Commun. Surv. Tutorials **18**(3), 2084–2123 (2016)
63. Huaqun, G., Xingjie, Y.: A survey on blockchain technology and its security. Blockchain: Res. Appl. **3**(2), art. 10072 (2022)
64. Sümer, B.: Can Self-Sovereign Identity (SSI) fit within the GDPR?: a conceptual data protection analysis (part I) (2022). https://www.law.kuleuven.be/citip/blog/can-self-sovereign-identity-ssi-fit-within-the-gdpr-part-i/
65. Sümer, B.: Can Self-Sovereign Identity (SSI) fit within the GDPR?: a conceptual data protection analysis (part II) (2022). https://www.law.kuleuven.be/citip/blog/can-self-sovereign-identity-ssi-fit-within-the-gdpr-part-ii/
66. Sümer, B., Schroers, J.: The new digital identity regulation proposal and the EU data protection regime (2021). https://www.law.kuleuven.be/citip/blog/the-new-digital-identity-regulation-proposal/

Solutions to Data Protection Challenges in Distributed Ledger and Blockchain Technologies: A Combined Legal and Technical Approach

Danaja Fabčič Povše, Alfredo Favenza, Davide Frey, Zoltán Ádám Mann, Angel Palomares, Lorenzo Piatti, and Jessica Schroers

Abstract Blockchains and distributed ledgers have attracted increasing attention since the introduction of the Bitcoin blockchain. The ability to run decentralized computations on open networks, on Bitcoin and on the Ethereum Virtual Machine, has led practitioners and researchers to investigate the use of blockchains and distributed ledgers for a variety of applications that involve the management of personal data. However, the very characteristics of such distributed ledger technologies (DLTs)—immutability, decentralization, and automation—appear at odds with data protection legislation like the European Union's General Data Protection Regulation (GDPR). This poses significant challenges when designing

D. F. Povše
Vrije Universiteit Brussels, Belgium
e-mail: Danaja.Fabcic.Povse@vub.be

A. Favenza
LINKS Foundation, Turin, Italy
e-mail: alfredo.favenza@linksfoundation.com

D. Frey (✉)
University of Rennes, Inria, CNRS, IRISA, Rennes, France
e-mail: davide.frey@inria.fr

Z. Á. Mann
University of Amsterdam, Amsterdam, Netherlands
e-mail: z.a.mann@uva.nl

A. Palomares
Atos, Madrid, Spain
e-mail: angel.palomares@atos.net

L. Piatti
InfoCert, Rome, Italy
e-mail: lorenzo.piatti@infocert.it

J. Schroers
KU Leuven, Leuven, Belgium
e-mail: Jessica.Schroers@law.kuleuven.be

N. El Madhoun et al. (eds.), *Building Cybersecurity Applications with Blockchain and Smart Contracts*, Signals and Communication Technology,
https://doi.org/10.1007/978-3-031-50733-5_7

applications involving personal data. This chapter provides an analysis of possible solutions to these challenges, including results from the literature, proposals for new solutions, and a discussion of challenges that remain despite these solutions. In all cases, solutions require a combination of legal and technical contributions. For example, legal interpretations must take into account the decentralized and general-purpose nature of DLTs, while solutions like mutable ledgers and geographically aware storage may provide answers to some legal concerns.

Keywords Distributed ledger · DLT · Blockchain · Data protection · Privacy · GDPR

1 Introduction

Since their introduction in 2008 with the Bitcoin cryptocurrency [1], blockchains have found widespread use in a variety of online applications, often dealing with personal data [2, 3]. However, as observed in our companion chapter [4], the use of blockchain, blockchain-like data structures, and, more generally, distributed ledger technologies (DLTs) often comes at odds with the requirements set by data protection regulations like the General Data Protection Regulation (GDPR) of the European Union (EU) [5].

In our companion chapter [4], we focused on issues and challenges emerging from the application of the GDPR to DLTs. We found that significant challenges arise from three main properties of DLTs: immutability, decentralization, and automation. Immutability directly clashes with principles of the GDPR, like the "right to be forgotten." Decentralization makes it difficult to identify key legal roles such as a data controller, or data processor. Automation clashes with data subject rights, such as the right not to be subject to solely automated processing. Some of these challenges may be more or less critical depending on the type of DLT being considered. For example, identifying the data controller becomes easier in the case of a permissioned DLT managed by a company or set of companies. But many of the challenges remain regardless of the type of DLT.

Given the importance and difficulty of data protection in DLTs, this area has attracted considerable research [6]. In particular, several technical solutions have been proposed to improve data privacy, such as zero knowledge proofs, mixing services, and ring signatures. While these techniques can indeed contribute to better data privacy in certain situations, they do not fully alleviate the legal concerns in relation to data protection regulations. There is a general misalignment between the computer science literature, which proposes sophisticated technical solutions that nevertheless fail to ensure compliance with data protection regulations, and the legal literature, which analyzes the compliance with such regulations in different setups, but without in-depth coverage of the technical possibilities.

This chapter thoroughly discusses the solutions proposed in the literature and provides hints for future research both at a legal and at a technical level. Like in our companion chapter [4], we focus our attention on the GDPR and cover all its relevant provisions and on the different types of DLTs (public vs. private, permissioned vs. permissionless DLTs) while identifying the solutions to the challenges identified therein. Lastly, this chapter collects some of the most mature and active projects that are trying to leverage DLTs to deliver data processing and transfer use cases, applying some of the solutions we outline on the next pages.

We start by recalling the challenges we identified in our companion Chapter [4] in Sect. 2. Then, Sect. 3 analyzes potential solutions. Section 4 showcases some specific projects on data protection in DLTs, Sect. 5 discusses related work, and Sect. 6 concludes the chapter.

2 Summary of Challenges

We start by summarizing the findings described in our companion chapter [4]. Like there, we structure our discussion based on the three key properties that we identified as posing challenges for data protection: **immutability**, **decentralization**, and **automation**.

2.1 Challenges Resulting from the Immutability of DLTs

Immutability is often advertised as what provides the reliability and security of DLTs. Indeed, the ability to maintain tamper-proof information allows DLTs to serve as a reliable log-book for a variety of applications. However, problems start when the information being permanently stored on the ledger turns out to be personal data. Data subject rights like the **right to rectification** (art. 16 GDPR) or the **right to erasure** (art. 17 GDPR) become impossible to guarantee for data that is stored on a ledger. Moreover, it is currently unclear whether some data types like hash values effectively constitute personal data. But if this is the case, their continuous processing during the system's operation raises issues related with the rights *to restriction of processing* (art. 18 GDPR) and **to object** (art. 21 GDPR) as well with principles such as **accuracy** (art. 5 (d) GDPR), **storage limitation** (art. 5 (e) GDPR), **purpose limitation** (art. 5 (b)), and **data minimization**. Finally, immutability also makes it impossible for data subjects to withdraw consent if consent information is stored on the ledger.

2.2 Challenges Resulting from the Decentralization of DLTs

The GDPR was conceived with a client-server model in mind. It is therefore not surprising that the decentralized nature of DLTs comes at odds with requirements of

GDPR such as the identification of a data controller (**principle of accountability** of art. 5(2) and **general responsibility** under art. 24) or the need to identify in which country the data is being stored and processed. In addition, like for immutability, decentralization also interferes with the ability to obtain and manage consent.

2.3 Challenges Resulting from the Automation in DLTs

With respect to automation, we identified three potential reasons for which DLTs may clash with the GPDR. First, the GPDR stipulates the right not to be subject to fully automated decision-making. Second, automation tools like smart contracts may themselves be used for unlawful purposes. Finally, automated tools like smart contracts may cause the violation of several GDPR principles. These include, for example, the principle of *lawfulness, fairness, and transparency* and that of *confidentiality*. Automated tools may hide some of the data from end users, while malicious or erroneous implementations may result in data leaks. Other requirements, like the need for the controller to record processing activities (art. 30 GDPR) or to notify of data breaches (art. 33), may also be vulnerable to malicious and erroneous code.

3 Possible Solutions to the Challenges

This section discusses how the data protection challenges identified in our previous chapter [4] may be addressed in the context of DLTs. We follow the same structure as in that chapter [4], investigating possible solutions for challenges stemming from immutability, decentralization, and automation in DLTs. We discuss both legal and technical measures for addressing the challenges, paying also attention to the different types of DLTs as well as more complicated constructs based on DLTs, such as smart contracts or decentralized autonomous organizations (DAOs). Table 1 summarizes our findings.

3.1 Immutability

As explained in the previous chapter's section on immutability [4], the immutability of DLTs leads to difficulties with regard to various data protection principles (accuracy, storage limitation, purpose limitation, and data minimization) and data subject rights (right to rectification, erasure, restriction of processing, to object). Furthermore, immutability makes it difficult to support the withdrawal of consent and to comply with data protection by design and by default. All these issues result from the fact that, normally, in a DLT it is not possible to change or delete data of past transactions.

Table 1 Summary of challenges and solutions

Challenges		Solutions						
		Not using DLT	**Keeping data 'off-chain' in a database, with hash pointers**	**Encryption of data, deletion of encryption keys**	**Using private DLTs**	**Using mutable blockchain-like data structures**	**Using legal scope of interpretation**	**Other technical solutions**
Immutability	**Right to rectification**	☑	~ Unclear status of on-chain hash	~ Data only made inaccessible	~ Requires loose interpretation	☑	~ If accepted by courts	✖
Immutability	**Right to erasure**	☑	☑	☑	~ Requires loose interpretation	☑	~ If accepted by courts	✖
Immutability	**Right to object**	☑	☑	☑	~ Requires loose interpretation	☑	~ If accepted by courts	✖
Immutability	**Right to withdraw consent**	☑	☑	☑	~ Requires loose interpretation	☑	~ If accepted by courts	✖
Decentralization	**Identification of the data controller**	☑	✖	✖	~ Requires loose interpretation	✖	~ If accepted by courts	✖
Decentralization	**Identification of the data transfers**	☑	✖	✖	~ Requires loose interpretation	✖	~ If accepted by courts	Sharding
Decentralization	**Consent management**	☑	✖	✖	✖	✖	✖	SSI
Automation	**Right not to be subject to fully automated decisions**	☑	✖	✖	✖	✖	~ If accepted by courts	Auditing

☑ Solution solves the challenge
✖ Solution does not help
~ Solution solves the challenge with the specified caveat

To address this challenge, different solutions have been proposed: to keep the personal data off the chain, to use private or mutable DLTs, or to interpret the legislation differently. These different solutions are analyzed below.

3.1.1 Keeping Personal Data Off the Chain

The usually mentioned solution is to simply keep the personal data off the blockchain. Different variations of this approach have been proposed [7, 8], from not using DLT for personal data, over storing the personal data somewhere else and only including hash pointers on the DLT, to storing the data in encrypted form on the DLT and deleting the encryption keys if necessary. The advantages, disadvantages, and open questions with regard to these approaches are discussed next.

Not using DLT when personal data is processed Not using DLT to process personal data would in principle solve the issue of compliance with data protection legislation since, as no personal data are processed, data protection legislation does not apply. The open question that remains is whether it is possible to completely refrain from processing personal data with DLT.

Keeping the data "off-chain" in a database, with hash pointers on the DLT Another possible solution is to segregate personal data in a separate "off-chain" database and keep only a hash of this data on the blockchain as a pointer to the personal data [9]. This makes it possible to erase data in the external storage when someone exercises the right to be forgotten, making the referral information on the blockchain useless. DLT transactions would only contain information needed to access the personal data in the separate database. In this manner, it would be possible to confine personal data to the off-chain storage and avoid storing such data on the DLT [10]. Off-chain storage would enable the modification and erasure of personal data stored off-chain in appropriate databases in line with Articles 16 and 17 GDPR. Yet, as observed by [11] and the CNIL [8], hash pointers are pseudonymous. Moreover, deletion of an item may turn out to be impossible if a non-compliant third party makes a copy of the data being stored before this is deleted.

This solution would resolve the issues with regard to the right to erasure, right to restriction of processing, and right to object. It would also allow compliance with the principles of storage limitation, purpose limitation, and data minimization.

The right to rectification and with it the principle of accuracy might remain difficult. If a hash of the data is included on the DLT and the data in the database is rectified, the hash will change and will not be the same as on the DLT anymore. This would usually defy the purpose of including the hash in the DLT. Furthermore, as long as the hash is connected to the personal data, it is considered to be personal data itself. In principle, this should be no problem, if it loses this status as soon as the connected personal data are deleted. However, the status of the on-chain hash, after the data has been erased, has not been explicitly stated yet [7, p. 97].

Encrypting the data, deleting the encryption keys Another proposed solution is to encrypt all the personal data on the blockchain with a key that allows only the associated data subject, and possibly a few authorized parties, to access the data. If a data subject requests their blockchain data to be deleted, the key is deleted, making the information inaccessible. Simple hashing is not sufficient to make data no longer identifiable. Encryption can make data anonymous if the encryption is strong enough and the encryption key is deleted and cannot be restored by anyone. Not deleting the encryption key would mean the data could be decrypted, and hence the individual would be identifiable. Some authors [8, 12] have proposed storing only encrypted personal data on a ledger and handling erasure requests by throwing away the decryption key. Clearly, this key should not be stored on the DLT, or its deletion would become impossible.

3.1.2 Using Private DLTs

The use of private or enterprise blockchains represents another possibility for compliance with GDPR's requirements. Private blockchains could limit the dissemination of personal data to just one company or a limited number of companies in a predefined consortium [11]. This would in turn limit the access to sensitive information to only a few individuals, thereby significantly reducing the possibility of data breaches. A relatively loose interpretation of the GDPR may already permit the storage of personal data in permissioned blockchains without risking violations. A first possible such interpretation could consist in allowing "erasure" to consist in restricting access rights to a data subject's personal data so that only the data subject can view. A second interpretation could consist in classifying hashed personal data as anonymized data, even though this goes against current opinions [8, 11]. With these interpretations, private blockchains that store personal data via a hash function and/or by enabling adequate access restrictions would not violate the GDPR [13]. The European Blockchain Service Infrastructure (EBSI) is following a permissioned approach leveraging a private, Ethereum-based implementation (Hyperledger Besu).[1]

3.1.3 Using Mutable Blockchain-Like Data Structures

Another potential solution that has been suggested in the literature is the use of editable (aka mutable) blockchains. Given the definition we gave in this paper, the term "editable blockchain" may sound like an oxymoron: a blockchain is an "append-only data structure," thus making it editable goes against its definition. But apart from this philosophical remark, the idea of having data structures that share some properties of the blockchain but are mutable remains interesting.

[1] https://ec.europa.eu/digital-building-blocks/wikis/display/EBSIDOC/Ledger+API

The concept of a mutable blockchain-like data structure is inherently tied to that of private, consortium blockchains. As observed by Politou et al. [11, 14], permissionless DLTs use immutability as a means to establish trust. Since trust in a third party already exists in private (permissioned), DLTs makes immutability less critical and thus makes it easier to design mutable blockchain-like data structures. The first mutable blockchain was proposed by Ateniese et al. [15] and was the subject of an Accenture patent.[2] It uses chameleon hashes [16], which support a trapdoor that, once known, makes it possible to identify arbitrary collisions, thereby rendering the hash non-unique. However, the use of a trapdoor requires either the presence of a trusted third party that will decide when to use that trapdoor or a secret-sharing mechanism among a set of nodes. The authors warn that sharing the trapdoor key among a large set of nodes (>200) may lead to performance issues. As a result, they recommend sharing the trapdoor key among a preselected committee. However, this causes the permissionless setting to fall back onto a permissioned one, at least for the sake of editing the blockchain.

A more recent system, μChain [17], proposes a solution in which the blockchain maintains several encrypted versions of each transaction. This makes it possible to revert a transaction by switching the active version. This is done by hiding the key corresponding to the old active version and revealing the key corresponding to the new version. In this case, the decision to switch versions can be managed by the entire set of miners, making it possible to really support permissionless systems. However, it is not clear what happens to transaction versions that are hidden after a change, as the corresponding key may already be out in the wild.

Deuber et al. [18] highlight a further problem with μChain: the fact that a malicious user may not include alternative versions and is free to prevent others from reverting his transaction. They thus propose a scheme that addresses this issue while improving scalability by not relying on multiparty computation. Their approach effectively allows nodes to remove all information that is edited out of the blockchain without relying on encryption, but only on majority voting. As a result, it requires (like other blockchain properties) a majority of honest miners.

The scheme proposed by Deuber appears to provide enough flexibility to enable blockchain applications to remove data when needed, without significantly hampering the security properties of the blockchain. This should make it possible to address most, if not all, of the issues associated with immutability, with the possible exception of public keys (see the discussion in our previous chapter's section on private keys [4]). It remains to be seen whether this technology or its future evolutions will become prominent in the blockchain ecosystem.

[2] https://cointelegraph.com/news/accenture-secures-patent-for-its-editable-blockchain-technology

3.1.4 Using Legal Scope of Interpretation

Ambiguities in the interpretation of the legislation could in some cases be used to argue that certain issues could be solved by a different interpretation of legislative provisions. This is, however, only possible to a certain degree and will depend in the end upon the acceptance of the interpretation by supervisory authorities as well as the courts as the final legal instance.

Right to rectification If the data has been stored on the blockchain itself, it would not be possible to adjust the data itself, but only to add new data which corrects it [19, p. 24][8, p. 10]. If courts accept this solution, then this addendum could also be considered as informing the other nodes of the rectification. Therefore, this would also possibly satisfy article 19 GDPR, which requires the data controller to also inform other parties who are using the data.

Right to erasure If the personal data to be erased is stored in the blockchain, it cannot be simply erased. According to Finck, the law stipulates that controllers should take account of available technology and the cost of implementation, but this only relates to the obligation to inform other controllers who are processing the personal data, and therefore it is not clear whether this exception could be taken into account in the case of DLTs [19, p. 24].

Another possible approach is that the notion of "erasure" is not defined and could therefore possibly also be interpreted in a more lenient way, for example, only making data inaccessible or limiting the processing, if deletion is not possible [19, p. 25]. The ICO, the British data protection authority, refers to this as *putting data beyond use* [20]. Essentially, this means that data cannot be used to inform any decision in respect of any individual or in a manner that affects the individual in any way, that no other organization can access the data, that proper technical and organizational security measures are put in place, and that the controller commits to permanent deletion of the information when or if that becomes possible. However, since eventual deletion is not possible on DLTs, it is not certain that other data protection authorities would consider this method sufficient for compliance [19].

Another argumentation is based on art. 17 (1) (a) GDPR, which provides that the right to erasure can apply if the personal data are no longer necessary in relation to the purposes for which they were collected or processed. The argument in that case could be that the processing of the data as it is included in the DLT is still necessary either for the operation of the DLT or for other purposes to bez evaluated on a case-by-case basis. Similarly, Art. 17 (1) (b) GDPR provides that the right to erasure applies if the data subject withdraws consent. In that case, it could potentially be argued that the core functioning principle of a technology is a legal ground for the processing [9, p. 426]. However, these arguments would still need a careful analysis before the personal data is added to the DLT and would probably in many cases not be accepted.

Right to object, right to restriction of processing In some cases (see the section on immutability in [4]), the issue is that the regular verification of the whole blockchain could be considered processing of personal data, and such processing might not be allowed if the data subject invoked the right to object or the right to restriction of processing. For the right to restriction of processing, it seems reasonable that the inadvertent processing of the data without any other effect could be considered to fall under the exception of storage (art. 18 (2) GDPR). However, different from the right to restriction of processing, the right to object does not include a storage exception. It might be possible to invoke the legitimate interest of the controller to continue using the DLT, assuming that the balancing exercise is carried out appropriately.

3.2 Decentralization

The decentralization of actors conflicts with GDPR's notion of a controller as a central processing entity. The designation of the controller will depend on the type of ledger used: in the case of permissioned blockchain, whoever operates and controls the blockchain deployment may be the controller. However, the situation is less clear for permissionless ledgers, as there is no entity that "controls" the blockchain. The situation is exacerbated by the global nature of the DLTs, as nodes may be located anywhere in the world.

3.2.1 Determining the controller

The data controller is the entity that determines the purposes and means of the processing of personal data, alone or jointly with other entities (art. 4(7)). Ascertaining who determines the means and purposes of processing can be carried out in three ways under the applicable law.

Option 1: applying the household exemption The household exemption is specified in art. 2(2)(c), stating that the GDPR does not apply to the processing of personal data by a natural person in the course of a purely personal or household activity. This exception must however be interpreted restrictively and, for example, does not include filing systems intended to be used by other persons, as decided in the 2018 Jehovah's witnesses case [21].

In a blockchain context, lightweight nodes only perform operations by interacting with full nodes. As a result, they do not store transactions and can be considered to fall under the household exemption, especially when their operations only involve their own personal data [9]. Likewise, users who only submit their own personal data to the blockchain might not be considered as falling under the scope of the regulation as a result of the exemption [22]. For example, this is the case in self-sovereign identity solutions, where users (acting as holders) only process their own information.

Option 2: existing case law on joint controllership—the "effective means" test Recently, the opinions of the EDBP and the case law of the CJEU have focused on the "effective means" as the factor in determining the controller [7, 23].

Two recent decisions of the Court of Justice of the EU (CJEU) have dealt with the notion of (joint) controllership. In its ruling C-210/16, Unabhängiges Landeszentrum für Datenschutz Schleswig-Holstein v Wirtschaftsakademie Schleswig-Holstein GmbH, (hereafter: the Wirt-schaft-akademie decision), the Wirtschaftsakademie was running a Facebook fan page without notifying visitors that their information was being collected through cookies. The court, following the Advocate General's opinion, ruled that the determination of a controller is factual rather than formal and must be interpreted broadly in order to ensure the effective and complete protection of data subjects.

In the second decision, C-40/17, Fashion ID GmbH & Co. KG v Verbraucherzentrale NRW eV (hereafter: the Fashion ID decision), the online retailer Fashion ID embedded a Facebook Like button on its website, through which personal data were transferred to Facebook without the data subject's knowledge. When a consumer protection organization brought proceedings against it, the retailer claimed it had no control over data transmission to Facebook nor how the latter would use those personal data. The court ruled that the embedding of a Facebook Like button points to a decisive influence of Fashion ID on the data transfer to Facebook, which would not have occurred without the plugin. In other words, "the fact that Fashion ID does not have access to the data collected and transmitted to Facebook did not change CJEU's conclusion" [24] that Facebook and Fashion ID are joint controllers.

Keeping the "effective means" test in mind, that means that in DLTs there are the following options for a controller:

- Each full node/miner separately is a controller. While lightweight nodes probably fall under the household exemption (see Option 1 above), full nodes, who create and store transfers, do not as they provide essential contribution. Nevertheless, it remains difficult to see full nodes as controllers, since they themselves cannot determine the means and purposes of processing. Similarly, miners, who collect data into blocks and validate it, make essential contributions but are unable to determine the means and purposes of processing on their own [25].
- Another option is to consider the collective of full nodes, who hold the economic power, and miners, who can provide processing power. Together, they can modify or break the consensus protocol by making changes called forks. Soft and hard forks in the code are users who have the power to adopt new rules and thus have the power to determine the why and how of processing, thus potentially falling under the definition of the controller [25].
- Network users who submit personal data as part of their business activities might be considered controllers insofar as they are not transferring their own personal data [22].

All three of these possibilities are characterized by either collective controllership (the entire network as one controller) or joint controllers. However, enforcing the law in an environment with potentially hundreds or thousands of controllers would be impossible—instead, a third argument can be made that identifying a controller should be based on a contextual assessment.

Option 3: contextual assessment rather than upfront designation of a controller Since an upfront designation of a controller does not seem feasible in many cases, DLTs can instead be seen as a general-purpose technology. As an analogy, a general controller cannot be determined for the Internet as a general type of technology. Instead, controllers are defined for specific applications (e.g., website owners are controllers for their website). The same argument could be applied to DLTs [26]. Thus, the entity running the blockchain (i.e., providing means of processing) as the responsible party for implementing measures such as privacy by design could, by implication, be considered a controller. This approach, however, requires identifying the "entity running the blockchain."

To do so, parallels can be drawn to the entities running the Internet as another general-purpose technology:

- Nodes and Internet intermediaries (e.g., ISPs, cloud service providers) can act as either data controllers or data processors, depending on whether they process data for their own ends [23, 27]. Nodes that process data on the blockchain for users' purposes and not their own could likewise be considered processors and not controllers.
- Hosting data on (external) servers: following the EDPB's argument in its guidelines on controllership that, while purely hosting personal data corresponds to being a processor, if the host can do anything further with the data, then it should be considered a data controller [27].

3.2.2 Transfer of Data Outside the EU

As analyzed in the section on challenges of decentralization in [4], decentralization leads to further challenges when controllers are based in non-EEA jurisdictions. While private permissioned and public permissioned DLTs can carefully select nodes and/or clients based on their geographical location, public permissionless DLTs remain the problem due to the lack of such control. In [4] we described the four options under the GDPR for transfers to non-EEA jurisdictions: (a) territory with an adequacy decision, (b) transfer under additional safeguards, (c) binding corporate rules, and (d) exception for non-repetitive transfers, if explicit consent is given.

We identify three ways to tackle compliance of global DLTs. The first two have a legal flavor: either finding a legal basis for international transfers under the GDPR or political action which could lead to regulatory convergence. The third relies on technical advancements that are currently being investigated.

Solutions under the GDPR (de lege lata) Potential solutions depend on the type of DLT:

- *Private permissioned DLT*: the company running the blockchain has control over the location of nodes as well as clients. Since the company is likely to be considered the controller, it will need to ensure adherence to the GDPR rules on potential international transfers. Clearly, it can opt for either option (a) or (b). This implies careful selection of the nodes, taking into account their geographical location, similarly to choosing a data processor. Following art. 28(1), the controller can only use those processors that provide sufficient guarantees to implement appropriate technical and organizational measures in such a manner that processing will comply with the rules of the GDPR and ensure the exercise of data subject rights. If the company running the blockchain belongs to a group of undertakings, binding corporate rules are also an option.
- *Public permissioned DLT*: Similarly to the private permissioned case, the consortium as the data controller can choose either option (a) or (b). Depending on the status of the clients as either controllers or processors, additional agreements must be concluded: an arrangement with division of responsibilities among the joint controllers, especially on the exercise of data subjects' rights (art. 26 of the GDPR), or a controller-processor agreement under art. 28(3). Binding corporate rules under (c) can only be an option insofar as the clients and the company/consortium can be considered to belong to the same group. However, it is currently technically impossible to control where nodes are located, even if the chain's protocol could in theory be amended to restrict processing to EEA-based nodes [19].
- *Public permissionless DLT*: this case does not seem to fit any of the four options.

Option (a) could only be relevant if all jurisdictions in the world had an adequacy decision, which is politically unlikely to happen.

Adopting appropriate safeguards under option (b) is potentially relevant, but could be difficult in practice. The first step in ensuring compliance of international transfers is mapping the personal data and knowing where it is located. This can be extremely challenging if a multitude of controllers and processors are involved. While the EDPB does not mention DLTs explicitly, they are definitely an example of such a complexity [28].

Option (c), binding corporate rules, is discussed by Renieris and Greenwood [29]. Transfers might be possible insofar as the network rules are legally binding, offer data subjects a mechanism to enforce their rights, and ensure the minimal data protection standards foreseen in art. 47(2). However, it is not clear how this would work in practice, considering how difficult it is to enforce data subject rights in public permissionless environments in general. Nor is it clear whether a public permissionless blockchain falls within the meaning of "group of undertakings."

The last remaining possibility to lawfully transfer data is the derogation contained in art. 49. Art. 49(1) allows for an exemption from options (a), (b), and (c) insofar as any of these options is not possible, but the transfer is only permissible if it meets restrictive criteria. First, the data subject has given explicit consent after being informed about the possible risks. Second, the transfer is not repetitive, it concerns only a limited number of data subjects, and it is necessary for the purposes of compelling legitimate interests pursued by the controller which are not overridden by the interests or rights and freedoms of the data subject. Finally, the controller has assessed all the circumstances surrounding the data transfer and has on the basis of that assessment provided suitable safeguards with regard to the protection of personal data. However, art. 49 is a derogation, which must be interpreted restrictively lest it become the rule [28].

It is difficult to tell whether transactions in a public permissionless blockchain meet all these criteria under option (d). While obtaining explicit consent to third-country transfer could be realistic, the question remains whether blockchain transactions fall under the non-repetitiveness criterion. The EDPB explains that non-repetitive transfers "may happen more than once, but not regularly" and this happens "outside the regular course of actions, for example, under random, unknown circumstances and within arbitrary time intervals" [28]. DLTs are persistent and decentralized, resulting in regular, constant transactions of personal data, which renders transfers based on the explicit consent derogation unlikely to be lawful.

No data protection authority has yet declared a public permissionless blockchain unlawful, despite the difficulties in finding a solid legal basis for international transfers. However, this does not mean that enforcement will never occur, resulting in a difficult situation for operators.

Regulatory convergence—potential legal solutions (de lege ferenda) Global regulatory harmonization could help mitigate risks relating to market functioning of DLTs [30]. However, that depends largely on political consensus: if DLTs are seen by policy-makers as the equivalent of the next Internet, convergence of regulatory guidelines is likely to follow, at least regarding permissioned blockchains (both public and private) [31]. On the other hand, pseudo data location requirements are not seen as desirable by policy-makers such as the European Commission. Nor are such requirements realistic, since DLTs (and the Internet) are global, and putting data location requirements in place could be a potentially large barrier for European companies who are active on a global scale [32].

Technological solutions Instead of trying to understand how to apply the law to a hard-to-control system, a possibility consists in building systems that naturally and by their very nature abide by the law. In the context of DLTs, the PriCLeSS project[3] is currently exploring whether sharding solutions can improve ledgers not only with respect to scalability but also by making them more legally compliant. In layperson terms, sharding consists in the decomposition of a ledger into multiple sub-ledgers (the shards) each responsible for a subset of the data and a subset of the transactions. It is therefore natural to see sharding as an opportunity to store data only on devices that are legally allowed to do so. For example, one could imagine sharding a ledger so that data from a company is only stored on devices managed by

[3] https://project.inria.fr/pricless/

partner companies and not competitors. Similarly, in an international context, one could imagine that data about EU citizens will remain on shards managed by nodes located within the EU.

Although this may sound like a simple and straightforward solution, the approach clashes with one of the principles that ensure the safety and security of distributed ledger: the unpredictability of storage locations. Sharding solutions leverage randomness in the assignment of nodes to shard to limit the possibility of an attacker to gain control of an entire shard. The challenge being addressed in PriCLeSS lies in balancing these two contrasting needs.

3.2.3 Consent Management in a Decentralized Environment

According to the GDPR, and as stated at the end of the section on the legal background of DLTs in [4], consent management involves three steps: collection, update, and revocation. Collection refers to consent being collected from the data subject and stored in compliance with the accountability principle. The update step occurs each time the privacy-notice use-consent collection is modified. When this happens, consent must be collected again. The revocation step happens when the data subject revokes their consent. This may happen at any time.

In a decentralized environment, and as far as a data controller is identified (according to Sect. 3.2.1), consent management may be an issue (as described at the end of the section on decentralization in [4]), but it can be managed thanks to different entry points: applications built on a blockchain may leverage smart front-ends that can interact with the data subject in order to collect consent.

In the case of online wallets, for example, the company providing the wallet—e.g., an exchange or gaming platform—appears the most likely data controller as discussed in Option 3 in Sect. 3.2.1. The company can clearly leverage the wallet for collecting consent: either by placing a special transaction on the wallet or simply by requesting a confirmation on the website before accessing the service. However, this may be a good approach for solving the collection part of the problem, but update and revocation remain difficult if the user can interact with the wallet outside the web-based platform.

A potential solution, which would however require some research, could be to leverage the feature of self-sovereign identity platforms. For example, we could imagine the data subject acting as an issuer of a credential that grants consent to a data controller, acting as a credential holder. Legal authorities would then take the role of verifiers and verify that the credentials held by the data controller are valid. This framework would, for example, allow the data subject to revoke an issued credential at any time. Optionally, smart contracts could further automate this process, by enforcing checks on the consent credentials held by data controllers.

3.3 Automation

We now discuss the solution to the challenges posed by automation. We divide our discussion into three parts. First, we consider the legal solutions, then technical solutions, and then apply them to the examples we gave at the end of the automation section in [4].

Legal Solutions As discussed in the section on automation in [4], solely automated data processing can pose significant challenges when it comes to a valid legal basis of data processing. Possible legal solutions can come from the specific exceptions to the general prohibition of solely automated data processing set in Article 22 of the GDPR.

The most common exception refers to the collection of consent, in the form of a compliant privacy notice. The controller should obtain explicit consent from the data subject for any automatic data treatment or decision based solely on automated processing that produces legal effects on the data subject. Consent can be obtained in a decentralized setting using the solutions outlined in Sect. 3.2.3. Another exception involves processing that is necessary for entering into, or executing, a contract between the data subject and a data controller. Finally, the third exception involves processing that legally authorized by the member state law to which the controller is subject.[4]

Technical Solutions From a technical perspective, a promising solution to the hurdles of automated decision-making comes from auditing. Auditing a smart contract—or a DAO, a form of a complex smart contract—can contribute to understanding and outlining it specific behavior. As a result, audited smart contracts allow all the stakeholders to fully understand the logic behind the automation, enabling a transparent data treatment. While auditing can be achieved manually, this is labor-intensive and error-prone. Automated auditing tools, albeit still in their infancy, offer a promising alternative.

In particular, a number of authors have proposed the use of formal verification techniques for auditing smart contracts. Murray and Anisi [33] propose a survey of existing approaches to the formal verification of smart contracts. In particular, they consider several contributions in the area of formal verification of smart contracts. One of the earliest such contributions [34] proposes rewriting smart contract using the F^* [35] language, to enable formal verification through a combination of SMT solving and manual proofs. However, their SOLIDITY to F* compiler (SOLIDITY*) cannot support many of the features of SOLIDITY, which limits its applicability. The paper also proposes EVM*, a decompiler which can translate Ethereum byte code into F*, making it possible to analyze smart contract with no publicly available source code. Similarly, KEVM [36] provides

[4] For more details, please refer to Article 22(2) of the GDPR.

an executable formalization of the Ethereum Virtual Machine based on the K Framework [37]. Another approach consist in using model checking. For example, the NuSMV [38] symbolic model checker was successfully used [39] to model the Ethereum blockchain, its smart-contract execution environments, and some smart contracts to demonstrate the feasibility of this approach.

Automated, or semi-automated, auditing is not only an interesting research topic, but also used by a number of companies. For example, OpenZeppelin[5] appears to use a semi-automated auditing technique with security experts that exploit tools to analyze smart contract code. Solidified[6] also performs semi-automated auditing thanks to a panel of experts that include cryptographers, distributed-systems researchers, and economists. They also offer code reviews and penetration testing. Diligence[7] is a product by Consensys that offer automatic scans as well as manual code reviews by security experts. They also advertise the ability to monitor the code for vulnerabilities while it's being maintained and changed. Finally, Solidity Finance[8] advertise a combination of automated tools, including static analysis, and manual code reviews.

Putting solutions into practice Let us now reconsider the examples we presented in the automation section of the previous chapter [4] in light of the solutions we just considered.

In the first example, a vendor uses a smart contract enriched with the data subject's personal data: applying what was set out just above, the data controller will need to comply with using one of the legal bases—or exceptions—under the GDPR and, above all, subjecting its code to audit, perhaps by a third party.

In the case of a DAO, the solutions outlined above need to be implemented with even greater care. It might be difficult to apply one of the exceptions, given the amount and variety of use of the tool, having to resort to consent. It remains, even in this example, necessary to audit the code, which might be more complex than the simple, single smart contract, mentioned in the first example.

In the third example, the applicable solutions remain the same, although things get more complicated: while the use of different DLTs does not change the approach to the legal basis, it will certainly be more complicated to perform a full audit. Indeed, each chain has its own computational logic that could increase the attack surface: in this case, it might be useful to perform more than one audit to ensure compliance.

---

[5] https://www.openzeppelin.com/security-audits

[6] https://www.solidified.io/#what-we-do

[7] https://consensys.net/diligence/

[8] https://solidity.finance/

3.4 Applying the Proposed Solutions to the Examples

3.4.1 Public Keys

As we discussed in the section on public keys in [4], it is often hard, if not impossible, to determine if a given public key should be considered personal data. Moreover, even when a public key can be determined to be personal data, the information that links it to an identified or identifiable natural person may not be directly available. From a data protection point of view, it may be best to react to this doubt by acting as if public keys consist of personal data. This would ensure protection of personal data and compliance with the legislation. However, from a technical perspective, it appears to be overkill and above all impractical to consider all the public keys stored on a blockchain as personal data. A more sensible approach would consist in considering the information concerning the link between a public key and a person as personal data. We argue that this measure could be sufficient to protect the privacy of individuals with respect to their public keys. In all cases, since perfect compliance with the GDPR does not appear to be possible in the case of public keys on the blockchain, a data subject using a blockchain should be aware of the potential consequences and reduced possibility to delete them later.

The difficulty to identify controllers also applies in the context of public keys, and actually the solutions we describe in Sect. 3.2.1 apply here. For example, if a data subject is using the blockchain and their public keys in the scope of their purely personal or household activities, the household exemption would apply. If, on the other hand, a person or company uses public keys of a data subject in a commercial transaction, then they would become the controller. Like for immutability, solutions that hide the link between a public key and the associated natural person, if any, appear to be the most promising.

When considering the risks associated with data transfer outside the EU, the solutions we considered in Sect. 3.2.2 also apply to public keys. In this respect, it is also worth noting that, since the transfer of public keys tends to be regular, it is unlikely that the derogation of art. 49 GDPR can be used.

With respect to the problem of linkability highlighted in the public keys section of [4], several technical solutions exist. One of them consists of deterministic wallets, which make it possible to derive many keys from a single "seed." The most advanced form of deterministic wallets is the HD wallet defined by the BIP-32 specification.[9] HD wallets contain keys derived in a tree structure, such that a parent key can derive a sequence of children keys, each of which can derive a sequence of grandchildren keys, and so on, to an infinite depth. Hierarchical deterministic wallets have also been proposed as a solution to privacy in private distributed ledgers. An HD wallet uses algorithms to create a new public-private key pair for each transaction or piece of a larger trade, generating a virtually infinite stream of keys from a single master

[9] https://github.com/bitcoin/bips/blob/master/bip-0032.mediawiki

seed. This can provide unlinkability and make the user's identity difficult to trace, provided that the user correctly refrains from reusing previous keys.

Another solution providing unlinkability consists in the use of ring signatures. A ring signature combines a group of individual signatures to produce a unique signature that can be used to trigger a transaction. One of the most widely known examples of a blockchain using ring signatures by default to protect privacy is Monero.[10] To make it hard to identify a transaction's sender, ring signatures combine his/her identity with that of other users, making it computationally infeasible to determine which one originally generated the transaction.

These solutions do not prevent public keys from being personal data, but they at least solve the linkability problem.

3.4.2 Wallet Solutions

To address the data protection challenges faced by DLT wallets, several possible solutions have been proposed. In order to secure the private key management, hardware wallets can significantly reduce the risk of unauthorized access or theft [40] Additionally, multi-signature wallets require multiple keys to authorize a transaction, providing an extra layer of security. The integration of privacy-enhancing technologies, such as zero-knowledge proofs, into blockchain wallets can help protect user anonymity and maintain transaction privacy while preserving the transparency of the underlying blockchain [41]. Another countermeasure involves the use of strong authentication methods, such as two-factor authentication (2FA) and biometric authentication, which can help protect user accounts and minimize unauthorized access [42].

3.4.3 Self-Sovereign Identity

In the section on self-sovereign identity in [4], we outlined the problems that may arise due to the use of a blockchain in SSI systems. In particular, we highlighted that even the storage of public DIDs may still raise issues with respect to the principle of data minimization and the right to erasure. The problem results from the immutable and persistent nature of the data that is stored on a blockchain. Once a DID has been published there, it cannot be unpublished.

The solution to this lies in assessing the actual requirements of DID storage in SSIs. It is true that public DIDs do need to remain publicly accessible. But they only need to do so for as long as they are active. Thus, a blockchain may not be the wisest choice to implement a DID registry. Projects like SOTERIA[11] (Sect. 4.5) are

---

[10] https://www.getmonero.org/

[11] https://www.soteria-h2020.eu/

currently working on blockchain-less solutions for self-sovereign identity systems, for example.

Another issue that was mentioned in the section on self-sovereign identity in [4] relates to the use of persistent identifiers in eIDAS and eIDAS2.[12] But, as already pointed out, this is not strictly an SSI problem, but it is associated with the way the eIDAS regulation interprets the concept of identity. From an SSI viewpoint, persistent identifiers are not a requirement.

4 Projects

This section gives some examples of recent projects using DLTs in conjunction with personal data processing.

4.1 DIZME

DIZME[13] is a trust framework based on the SOVRIN Foundation framework and ledger. The project is based on the self-sovereign identity (SSI) paradigm, which aims to give back to the identity owner (the "Owner") the management of his/her digital identity. The solution is technically based on ZKP and data minimization-oriented tools. Most of the pillars of the GDPR are also embraced by the SSI guiding principles:

- Control: users must control their identities
- Access: users must have access to their own data
- Transparency: systems and algorithms must be transparent
- Portability: information and services about identity must be transportable
- Consent: users must agree to the use of their identity
- Minimization: disclosure of claims must be minimized

The SOVRIN ecosystem is based on two W3C standards: Decentralized IDentifier ("DID") and Verifiable Credentials ("VC" or simply "Credential").

This approach provides for three main players:

- Issuer: the legal/natural person who issues the Credential
- Verifier: the legal/natural person that asks for and verifies the Credential
- Holder: the legal/natural person to whom the Credential is related

[12] https://digital-strategy.ec.europa.eu/en/policies/eidas-regulation

[13] The name is a contraction of "this is me." An early version of the website can be found under https://dizme.io/.

Once the Issuer has given the Credential to the Holder, the Holder can spend it in front of the Verifier for any needed purpose, without involving the Issuer. This approach helps to protect personal data because on one hand the Issuer does not know where the Holder is spending his/her credential, and on the other hand it allows the Holder to have—by design—most of the rights given by the GDPR (see the section on data subjects' rights in the legal introduction of [4]).

DIZME is a domain-specific governance framework (DSGF) for trusted identity, and it comes with a specific mobile application (DIZME wallet) that enables the control of the Owner over his/her digital identity. In the DIZME framework, Issuer, Verifier, and Holder interact easily with a QR code system that allows to ask for a specific Credential, which is seamlessly spent through the wallet itself. The wallet is bound to the Holder through a secure onboarding procedure: from this starting point, the user can start to ask Credential issuing, according to the level of assurance he/she needs. The user can choose three different levels of identification: level 1 (L1) is a self-assessed identity, level 2 (L2) is L1 corroborated by some ID checks, and level 3 is L2 corroborated with specific remote onboarding procedure compliance. This multi-level approach is in accordance with the data minimization principle. DIZME is also a bridge between the SSI and the Qualified Trust Services given in the eIDAS Regulation.[14]

4.2 KRAKEN

The KRAKEN project aims to enable the sharing, brokerage, and trading of personal data including sensitive data by returning control to both data subjects and data providers throughout the entire data lifecycle. The project is providing a data marketplace which allows the sharing of personal data and its usage for research and business purposes, by using privacy-preserving cryptographic tools.

To achieve this goal, KRAKEN is developing an advanced platform to share certified information between users and organizations by leveraging DLT, promoting the vision of self-sovereign identity solutions, preserving security, privacy, and the protection of personal data in compliance with EU regulations. The development carried out by the project is set out in three main pillars:

- Developing an SSI solution to provide a decentralized user-centric approach to personal data sharing.
- Implementing a set of analytic techniques based on advanced cryptographic tools to permit privacy-preserving data analysis.
- Integrating the above techniques into a data marketplace, allowing the sharing of personal data and privacy-preserving data analytics.

[14] "Regulation (EU) No 910/2014 of the European Parliament and of the Council of 23 July 2014 on electronic identification and trust services for electronic transactions in the internal market and repealing Directive 1999/93/EC"

The cryptographic tools provided by KRAKEN are based on secure multi-party computation (SMPC). SMPC allows a group of nodes to compute on secret inputs jointly, without disclosing their respective inputs to the other nodes or any other party. Even if one of the nodes is malicious, SMPC guarantees that this malicious node cannot infer anything about the input of the other nodes of the network. Furthermore, the correctness of the computation can be guaranteed as long as a sufficiently large fraction of the nodes behave honestly.

In addition, KRAKEN implements an identity and access management (IAM) approach, based on blockchain, for managing the identity of end users in order to empower data subjects to control their data. This SSI approach provides a solution where the end user has the whole control of their identity information in their own mobile device. In KRAKEN, the development of an SSI solution is going one step further by addressing one of the biggest challenges of SSI: what happens if the end user loses his/her mobile phone or has different devices where they want to use their identity information. KRAKEN implements a backup service of the different secrets and identity information necessary for the use of an SSI solution using cryptographic proxy re-encryption techniques that ensures that the secrets and identity information are never disclosed outside the end user's mobile device. KRAKEN also provides the functionality to obtain eIDAS compliance credentials and adheres to the recommendations and suggestions proposed by the EBSI and eSSIF bodies.

All these techniques and functionalities are applied by a marketplace, leveraging the privacy preservation of personal or even sensitive data. The marketplace is piloted in two different domains: eHealth and education.

4.3 European Blockchain Service Infrastructure: EBSI

The European Blockchain Service Infrastructure (EBSI) is a blockchain-based initiative launched by the European Union in 2019 to create a standardized and interoperable infrastructure for the delivery of cross-border public services. The EBSI is built on top of existing national blockchain networks and leverages open-source technology to enable secure and transparent data exchange across European borders. EBSI leverages leading standards such as the W3C Decentralized Identifiers (DIDs), Verifiable Credentials (VCs), and Verifiable Presentations (VPs), as well as OpenID for Verifiable Credentials, the General Data Protection Regulation (GDPR), the electronic Identification, Authentication, and Trust Services (eIDAS) regulation, and other relevant EU regulations. By building on these standards, EBSI aims to establish a generic profile for the entire lifecycle of self-sovereign identity (SSI), from credential issuance to verification and presentation. This approach enables individuals to retain control over their personal data and facilitates the secure and privacy-preserving exchange of verified information across different services and applications.

4.4 *PriCLeSS*

PriCLeSS (Privacy Conscious Legally Sound blockchain Storage[15]) is a French project funded by the CominLabs LabEx.[16] PriCLeSS consists of a partnership between computer-science and law researchers and aims at bridging the gap between GDPR and blockchain-based storage applications. The project addresses three major challenges. The first consists in leveraging the characteristics of distributed ledgers as a tool to automate the auditing of operations on personal data. The second consists in providing novel ledger designs that can take into account the requirements of GDPR like the enforcement of European borders in the context of data transfer. The third involves the design of an ecosystem of tools that can complete the blockchain with other resilient distributed data structures that can offer features that are currently absent in the blockchain context, like mutability, the ability to erase data, and access control.

4.5 *SOTERIA*

SOTERIA[17] is an H2020 Innovation Action led by IDNow, a leading identity-proofing provider in Europe. SOTERIA aims to build a decentralized platform for the management and storage of personal data, combined with advanced identity management tools. This platform will be tested in three use cases involving citizens. But beyond this high TRL endeavor, SOTERIA is also exploring the design of novel blockchain-less decentralized identity management solutions.

5 Related Work

Previous work on data protection aspects of DLTs consists of papers from a technical point of view (reviewed in Sect. 5.1) and papers from a legal point of view (reviewed in Sect. 5.2).

5.1 *Technical Literature*

Halpin and Piekarska provide a high-level overview of the security and privacy challenges of the blockchain [43]. The identified challenges include a lack of formally stated privacy and security properties, the difficulty of upgrading the

[15] https://project.inria.fr/pricless/

[16] https://cominlabs.inria.fr/

[17] https://www.soteria-h2020.eu/

cryptographic primitives used in a blockchain system, and the limited privacy and anonymity offered by blockchain. For the latter challenge, also some potential solutions—like mixing services or Succinct Non-interactive ARguments of Knowledge (SNARKs)—are mentioned.

Bayle et al. address the problem of how the blockchain may be used in such a way that it complies with one of the provisions of the GDPR, the right to be forgotten, despite the immutability of the blockchain [44]. They suggest keeping any personal data off-chain and only use the blockchain for keeping track of actions affecting the data. This way, personal data can be erased if necessary.

Li et al. survey security challenges of blockchain systems, with a particular focus on cryptocurrencies [45]. They identify nine main security risks and review several real-world attacks on blockchain systems, as well as proposals for enhancing the security of blockchain systems. They do not specifically consider data protection. However, they mention "transaction privacy leakage," i.e., the possibility for an attacker to (partially) re-identify pseudonymized transaction data, as one of the risks.

Feng et al. present a survey of privacy challenges in blockchain systems, also with a focus on cryptocurrencies [46]. They identify two main challenges: identity privacy (i.e., no leakage of identity information about the participants) and transaction privacy (i.e., no leakage of transaction data, such as the transferred amount). After presenting possible attacks on privacy, they review techniques that have been proposed for privacy protection in blockchain: centralized and decentralized mixing services, as well as different cryptographic approaches including ring signatures, non-interactive zero-knowledge proofs (NIZK), zero-knowledge succinct non-interactive arguments of knowledge (zk-SNARK), and confidential transactions.

Bernabe et al. present a comprehensive survey of privacy challenges and privacy-preserving solutions for blockchain, with a particular focus on identity management [6]. They identify eleven challenges, one of which is compliance with data protection regulations like the GDPR. They review several privacy-preserving techniques and group them into four categories: smart contract/key management, identity data anonymization, transaction data anonymization, and on-chain data protection.

A common shortcoming of these papers is that they do not address compliance with the GDPR as a whole. Existing technical solutions may enhance privacy, but this does not guarantee compliance with the GDPR.

5.2 Legal Literature

The literature surveyed often points to the conflicted relationship between data protection law and distributed ledger technologies. Nevertheless, authors often clarify that DLTs in themselves are not necessarily always incompatible with the law. This is because blockchain, like the Internet, is a general-purpose technology, and thus in order to assess legal challenges and compliance, we need to look at the context in which the blockchain is being used and keep in mind that in itself using the blockchain is not data processing [22, 26].

The main clashes occur in the area of scope of application (both material and territorial); responsibility and accountability, including the exercise of data subject rights; and implementing compliance with the obligations of the GDPR. Legal consequences also differ depending on what type of DLT is used (private/public, permissioned/permissionless), as well as which technological features are at stake [47].

Blockchains operate globally, while the GDPR only applies within predetermined territories; it could possibly apply to blockchain actors established outside the EU under the territoriality principle under its art. 3. However, this could make enforcement difficult, since only the EU authorities are authorized to enforce the compliance with the GDPR. [9, 19] Material scope of application refers to the concept of personal data in the DLT context. It is unclear to which data exactly GDPR could apply. Most authors consider that information such as public keys, user credentials, their copies, and revocations could be considered personal data [9, 48]; some question whether private keys should also fall under this definition [8, 19, 49].

The notion of controller is replaced by system architecture and code, which makes the DLTs more reliable but less flexible and less accountable [50]. It is unclear whether one actor or the entire network should carry the brunt of the responsibility to comply with the law and apply technical and organizational measures, especially in the case of public networks where either no node or every node where data are technically processed is responsible [9, 25]. The "accountability gap" means that if there is no controller, the data subject rights lose effectiveness [9]. It is important to understand which actors or entities have the responsibility; otherwise it could lead to further problems with enforcement, enforcing data subject rights, and determining who should implement measures [25]. It has been suggested to adopt a micro vs. macro perspective and that users of blockchain platforms are likely to be deemed controllers, while the miners and nodes act as processors [10]. In the case of public blockchains, their "gatekeepers" are likely to be targeted by regulation [9], thus leading to an assumption that they are to be considered responsible.

DLT in itself is not necessarily always against data protection law, and DLT technologies can function as tools that can facilitate data protection compliance: cryptography, data portability, and integrity are some of the features that can serve this objective. This leads to a curious dichotomy: privacy by design (and compliance) is either implemented in the DLT by default, or there is an intrinsic clash between them [9].

While many authors have suggested solutions to the challenges posed by compliance, those solutions are partial and rarely take into account both technological and legal possibilities or differentiate the solutions based on the type of DLT used.

Berberich and Steiner [9] considered the challenges posed by privacy by design, the right to be forgotten, territorial scope of application, and whether data on the blockchain can be considered personal data. However, they only consider private and public DLTs, without a distinction between permissioned and permissionless.

Technical solutions are discussed in abstracto, and legal solutions are limited to the context of privacy by design under art. 25 of the GDPR. Bacon et al. focus only on distributed versus centralized approaches [47] without suggesting viable solutions to bridge the gap.

Some technical solutions for compliance already exist, such as mechanisms that allow data deletion—even though a link would remain on the block, it might suffice to comply with a data erasure request. Likewise, deleting all instances of a private key could be considered sufficient. Technical advances in blockchain deletion might prove to be useful in implementing privacy by design obligation. [10] Other specific solutions could be, e.g., explicit consent under art. 49(1)(a), off-chain storage of PD to comply with the data minimization principle, and anonymization or shredding provide possible means of escape from the GDPR's scope [19, 51].

6 Conclusions and Outlook

This chapter presented a survey of the potential solutions for data protection in DLTs, together with some examples and projects studying this problem. In our companion chapter [4], we observed how the properties of immutability, decentralization, and automation make it difficult to operate DLT-based applications that comply with the GDPR. Here we analyze existing and potential solutions both from a legal and a technical perspective.

The considered solutions cover most of the problems in the case of private and permissioned DLTs, but some challenges remain, particularly in the permissionless case, where it is often difficult to identify a data controller. Ongoing projects are proposing novel technical solutions that can combine the benefits of distributed ledger with more flexible interfaces. At the same time, legal researchers and officials will play a crucial role in understanding how to apply existing regulations or develop new ones in the context of a dynamic technological environment.

References

1. Nakamoto, S.: Bitcoin: A peer-to-peer electronic cash system (2008). https://git.dhimmel.com/bitcoin-whitepaper/
2. Panetta, R., Cristofaro, L.: A closer look at the EU-funded My Health My Data project. In: Digital Health Legal pp. 10–11 (2017). https://doi.org/10.5281/zenodo.1048999
3. Zyskind, G., Nathan, O., Pentland, A.S.: Decentralizing privacy: using blockchain to protect personal data. In: 2015 IEEE Security and Privacy Workshops. pp. 180–184 (2015). https://doi.org/10.1109/SPW.2015.27
4. Fabčič Povše, D., Favenza, A., Frey, D., Mann, Z.Á., Palomares, A., Piatti, L., Schroers, J.: Data protection challenges in distributed ledger and blockchain technologies: a combined legal and technical analysis. In: El Madhoun, N., Dionysiou, I., Bertin, E. (eds.) Building Cybersecurity Applications with Blockchain Technology and Smart Contracts. Springer, Heidelberg, Germany (2024)

5. Timan, T., Mann, Z.: Data protection in the era of artificial intelligence: trends, existing solutions and recommendations for privacy-preserving technologies. In: The Elements of Big Data Value: Foundations of the Research and Innovation Ecosystem, pp. 153–175. Springer, Heidelberg, Germany (2021)
6. Bernabe, J.B., Canovas, J.L., Hernandez-Ramos, J.L., Moreno, R.T., Skarmeta, A.: Privacy-preserving solutions for blockchain: review and challenges. IEEE Access **7**, 164908–164940 (2019)
7. European Parliamentary Research Service: Blockchain and the General Data Protection Regulation (2019). https://www.europarl.europa.eu/RegData/etudes/STUD/2019/634445/EPRS_STU(2019)634445_EN.pdf
8. Commission Nationale Informatique & Libertés: Blockchain – solutions for a responsible use of the blockchain in the context of personal data (2018). https://www.cnil.fr/sites/default/files/atoms/files/blockchain_en.pdf
9. Berberich, M., Steiner, M.: Blockchain technology and the GDPR – how to reconcile privacy and distributed ledgers. Eur. Data Protection Law Rev. **2**(3), 422–426 (2016)
10. Bacon, J., Michels, J.D., Millard, C., Singh, J.: Blockchain demystified: a technical and legal introduction to distributed and centralized ledgers. Richmond J. Law Technol. **25**, 1 (2018)
11. Politou, E., Casino, F., Alepis, E., Patsakis, C.: Blockchain mutability: challenges and proposed solutions. IEEE Trans. Emerg. Top. Comput. **9**(4), 1972–1986 (2021)
12. Jensen, G.: Reconciling GDPR rights to erasure and rectification of personal data with blockchain. Oracle Cloud Security (2018). https://blogs.oracle.com/cloudsecurity/reconciling-gdpr-rights-to-erasure-and-rectification-of-personal-data-with-blockchain
13. Mirchandani, A.: The GDPR-blockchain paradox: exempting permissioned blockchains from the GDPR. Fordham Intellectual Property Media Entertain. Law J. **29**(4), 1201–1241 (2019)
14. Politou, E., Alepis, E., Virvou, M., Patsakis, C.: Privacy and Data Protection Challenges in the Distributed Era. Springer, Heidelberg, Germany (2022)
15. Ateniese, G., Magri, B., Venturi, D., Andrade, E.: Redactable blockchain–or–rewriting history in bitcoin and friends. In: 2017 IEEE European Symposium on Security and Privacy (EuroS&P). pp. 111–126. IEEE (2017)
16. Krawczyk, H., Rabin, T.: Chameleon signatures. In: Proceedings of the Network and Distributed Systems Security Symposium. pp. 143–154 (2000)
17. Puddu, I., Dmitrienko, A., Capkun, S.: μChain: how to forget without hard forks. IACR Cryptology ePrint Archive, 2017/106 (2017)
18. Deuber, D., Magri, B., Thyagarajan, S.A.K.: Redactable blockchain in the permissionless setting. In: IEEE Symposium on Security and Privacy (SP). pp. 124–138 (2019)
19. Finck, M.: Blockchain and data protection in the European Union. Max Planck Institute for Innovation & Competition Research Paper No. 18-01 (2017)
20. Information Commissioner's Office: Deleting personal data, guidance, 26.02.2014 (2014). https://ico.org.uk/media/for-organisations/documents/1475/deleting_personal_data.pdf
21. CJEU: Tietosuojavaltuutettu vs. Jehovan todistajat – uskonnollinen yhdyskunta. ECLI:EU:C:2018:551/C-25/17 (2018). https://curia.europa.eu/juris/document/document.jsf?docid=203822
22. European Union blockchain observatory & forum: blockchain and the GDPR (2018). https://www.eublockchainforum.eu/sites/default/files/reports/20181016_report_gdpr.pdf
23. Article 29 Data Protection Working Party: opinion 1/2010 on the concepts of "controller" and "processor" (2010). https://ec.europa.eu/justice/article-29/documentation/opinion-recommendation/files/2010/wp169_en.pdf
24. CJEU: Fashion ID GmbH & Co. KG vs. Verbraucherzentrale NRW eV. ECLI:EU:C:2019:629/C-40/17 (2019). https://curia.europa.eu/juris/liste.jsf?num=C-40/17
25. Buocz, T., Ehrke-Rabel, T., Hödl, E., Eisenberger, I.: Bitcoin and the GDPR: allocating responsibility in distributed networks. Comput. Law Secur. Rev. **35**(2), 182–198 (2019)
26. Moerel, L.: Blockchain & data protection… and why they are not on a collision course. Eur. Rev. Private Law **26**(6), 825–851 (2018)

27. European Data Protection Board: Guidelines 07/2020 on the concepts of controller and processor in the GDPR, version 2.0 (2021). https://edpb.europa.eu/system/files/2021-07/eppb_guidelines_202007_controllerprocessor_final_en.pdf
28. European Data Protection Board: Recommendations 01/2020 on measures that supplement transfer tools to ensure compliance with the EU level of protection of personal data (2020). https://edpb.europa.eu/sites/default/files/consultation/edpb_recommendations_202001_supplementarymeasurestransferstools_en.pdf
29. Renieris, E., Greenwood, D.: Unblocking blockchain data flows in the wake of Schrems II. MIT Computational Law Report (2020). https://law.mit.edu/pub/unblockingblockchaindataflowsinthewakeofschremsii
30. European Union blockchain observatory & forum: legal and regulatory framework of blockchains and smart contracts (2019). https://www.eublockchainforum.eu/sites/default/files/reports/report_legal_v1.0.pdf
31. Melin, K.: The GDPR compliance of blockchain: A qualitative study on regulating innovative technology. Thesis, University of Uppsala (2019)
32. Christakis, T.: After Schrems II: Uncertainties on the legal basis for data transfers and constitutional implications for Europe. European Law Blog (2020). https://europeanlawblog.eu/2020/07/21/after-schrems-ii-uncertainties-on-the-legal-basis-for-data-transfers-and-constitutional-implications-for-europe/
33. Murray, Y., Anisi, D.A.: Survey of formal verification methods for smart contracts on blockchain. In: 2019 10th IFIP International Conference on New Technologies, Mobility and Security (NTMS). pp. 1–6 (2019). https://doi.org/10.1109/NTMS.2019.8763832
34. Bhargavan, K., Delignat-Lavaud, A., Fournet, C., Gollamudi, A., Gonthier, G., Kobeissi, N., Kulatova, N., Rastogi, A., Sibut-Pinote, T., Swamy, N., Zanella-Béguelin, S.: Formal verification of smart contracts: short paper. In: Proceedings of the 2016 ACM Workshop on Programming Languages and Analysis for Security. p. 91–96 (2016)
35. Swamy, N., Hriţcu, C., Keller, C., Rastogi, A., Delignat-Lavaud, A., Forest, S., Bhargavan, K., Fournet, C., Strub, P.Y., Kohlweiss, M., Zinzindohoue, J.K., Zanella-Béguelin, S.: Dependent types and multi-monadic effects in f*. SIGPLAN Not. **51**(1), 256–270 (2016). https://doi.org/10.1145/2914770.2837655
36. Hildenbrandt, E., Saxena, M., Rodrigues, N., Zhu, X., Daian, P., Guth, D., Moore, B., Park, D., Zhang, Y., Stefanescu, A., Rosu, G.: Kevm: A complete formal semantics of the ethereum virtual machine. In: 2018 IEEE 31st Computer Security Foundations Symposium (CSF). pp. 204–217. IEEE Computer Society, Los Alamitos, CA, USA (2018). https://doi.org/10.1109/CSF.2018.00022, https://doi.ieeecomputersociety.org/10.1109/CSF.2018.00022
37. Roşu, G., Şerbănută, T.F.: An overview of the k semantic framework. J. Logic Algebraic Program. **79**(6), 397–434 (2010). https://doi.org/10.1016/j.jlap.2010.03.012, https://www.sciencedirect.com/science/article/pii/S1567832610000160. Membrane computing and programming
38. Cimatti, A., Clarke, E., Giunchiglia, E., Giunchiglia, F., Pistore, M., Roveri, M., Sebastiani, R., Tacchella, A.: NuSMV Version 2: An OpenSource Tool for Symbolic Model Checking. In: Proceedings of International Conference on Computer-Aided Verification (CAV 2002). LNCS, vol. 2404. Springer, Copenhagen, Denmark (2002)
39. Nehaï, Z., Piriou, P.Y., Daumas, F.: Model-checking of smart contracts. In: 2018 IEEE International Conference on Internet of Things (iThings) and IEEE Green Computing and Communications (GreenCom) and IEEE Cyber, Physical and Social Computing (CPSCom) and IEEE Smart Data (SmartData). pp. 980–987 (2018). https://doi.org/10.1109/Cybermatics_2018.2018.00185
40. Lehto, N., Halunen, K., Latvala, O.M., Karinsalo, A., Salonen, J.: CryptoVault – a secure hardware wallet for decentralized key management. In: IEEE International Conference on Omni-Layer Intelligent Systems (COINS) pp. 1–4 (2021)
41. Babel, S.: Bringing data minimization to digital wallets at scale with general-purpose zero-knowledge proofs (2023). arXiv:2301.00823

42. Zhang, Li, L.: Distributed blockchain-based data protection framework for modern power systems against cyber-physical attacks. IEEE Trans. Smart Grid **11**(4), 3130–3142 (2020)
43. Halpin, H., Piekarska, M.: Introduction to security and privacy on the blockchain. In: 2017 IEEE European Symposium on Security and Privacy Workshops (EuroS&PW). IEEE (2017)
44. Bayle, A., Koscina, M., Manset, D., Perez-Kempner, O.: When blockchain meets the right to be forgotten: technology versus law in the healthcare industry. In: 2018 IEEE/WIC/ACM International Conference on Web Intelligence (WI). pp. 788–792. IEEE (2018)
45. Li, X., Jiang, P., Chen, T., Luo, X., Wen, Q.: A survey on the security of blockchain systems. Futur. Gener. Comput. Syst. **107**, 841–853 (2020)
46. Feng, Q., He, D., Zeadally, S., Khan, M.K., Kumar, N.: A survey on privacy protection in blockchain system. J. Netw. Comput. Appl. **126**, 45–58 (2019)
47. Bacon, J., Michels, J.D., Millard, C., Singh, J.: Blockchain demystified. Queen Mary University of London, School of Law Legal Studies Research Paper no. 268 (2017)
48. Kondova, G., Erbguth, J.: Self-sovereign identity on public blockchains and the GDPR. In: Proceedings of the 35th Annual ACM Symposium on Applied Computing. pp. 342–345 (2020)
49. Manteghi, M.: Blockchain and the European Union's General Data Protection Regulation: from conflict to "peaceful" coexistence? (2021). http://dx.doi.org/10.2139/ssrn.3805647
50. Tatar, U., Gokce, Y., Nussbaum, B.: Law versus technology: blockchain, GDPR, and tough tradeoffs. Comput. Law Secur. Rev. **38**, Art. 105454 (2020)
51. Finck, M.: Blockchain and data protection in the European Union. Eur. Data Protection Law Rev. **4**(1), 17–35 (2018)

Index

N. El Madhoun et al. (eds.), *Building Cybersecurity Applications with Blockchain and Smart Contracts*, Signals and Communication Technology,
https://doi.org/10.1007/978-3-031-50733-5

SPRINGER NATURE

GPSR Compliance

The European Union's (EU) General Product Safety Regulation (GPSR) is a set of rules that requires consumer products to be safe and our obligations to ensure this.

If you have any concerns about our products, you can contact us on ProductSafety@springernature.com

In case Publisher is established outside the EU, the EU authorized representative is:

Springer Nature Customer Service Center GmbH
Europaplatz 3
69115 Heidelberg, Germany

The manufacturer's authorised representative in the EU is Springer Nature Customer Service Centre GmbH, Europaplatz 3, 69115 Heidelberg, Germany. If you have any concerns regarding our products, please contact ProductSafety@springernature.com

Printed and bound by CPI Group (UK) Ltd, Croydon, CR0 4YY
28/09/2026
02224086-0001